The
Great
Energy
Scam

Other books by Fred J. Cook

The Unfinished Story of Alger Hiss
What Manner of Men: Forgotten Heroes of the Revolution
A Two-Dollar Bet Means Murder
The Warfare State
The FBI Nobody Knows
The Corrupted Land
The Secret Rulers
The Plot Against the Patient
The Nightmare Decade, the Era of Joe McCarthy
Julia's Story: The Tragedy of an Unnecessary Death

Private Billions vs. Public Good

by Fred J. Cook

The Great Energy Scam

Macmillan Publishing Co., Inc.

New York

Macmillan Publishing Co., Inc.
866 Third Avenue, New York, N.Y.   10022
Collier Macmillan Canada, Inc.

Library of Congress Cataloging in Publication Data
Cook, Fred J.
    The great energy scam.
    Includes index.
    1. Petroleum industry and trade—United States. 2. Petroleum
industry and trade—Government policy—United States. I. Title.
HD9565.C5948   1982   338.2′728′0973   82-17076
ISBN 0-02-527800-2

10 9 8 7 6 5 4 3 2 1

Designed by Jack Meserole

Printed in the United States of America

To the dedicated men and women in the
trenches—those on Congressional and departmental
staffs—who do the real work of American
government and strive to keep it honest.

Contents

Acknowledgments

I cannot hope to thank all of those who helped me with this book and so made it possible. Some, for the sake of their careers, have to remain anonymous, but they know who they are and I am grateful to them. On another level, this book could not have been written without the support given me by Victor S. Navasky, editor of *The Nation*, its publisher and staff. Nor could it have been written without the support of Howard Bray, director of the Fund for Investigative Journalism, and the grants the fund made to cover some of my expenses, making it possible for me to be in the right places at the right times.

The
Great
Energy
Scam

Lifting the Lid

1

This is the story of a three-year investigation undertaken in an effort to find out what the major oil companies are doing to the American economy and the American consumer. When I began, I had no idea that, as an investigative reporter, I would devote so much time to trying to sift truth from bunkum. In the atmosphere now created by the Reagan administration, which holds in essence that we should leave all major decisions affecting our lives to the wisdom of big business, it becomes more essential than ever to expose the extent of the bunkum.

Two developments started my journalistic juices flowing, sluggishly at first, then with greater determination—but still with no conception that I was starting on a long and intricate search. In June 1976, I heard over my car radio that the Ford administration had persuaded Congress to decontrol all No. 2 home heating oil. I had no illusions about what this would mean to the American homeowner. In 1966 I had written the book *The Corrupted Land*, a large section of which had dealt with the great electrical price-fixing conspiracy first exposed by the late Senator Estes Kefauver. The case had involved major executives of such prestigious American manufacturers of electrical equipment as General Electric and Westinghouse. Powerful businessmen had run around using aliases like members of the Mafia; they had held whispered conferences in men's rooms, had met in isolated motels and had rigged the prices on electrical generators on which they were supposed to be bidding competitively. This major scandal of the time, now forgotten, and other incidents of a similar nature had convinced me that "free enterprise" is a myth perpetuated by the American business community, but that the reality is one of administered prices agreed upon by powerful potentates of multibillion-dollar businesses. And so my reaction to the news of heating oil decontrol was almost automatic. "The sons-of-bitches," I muttered to myself, "they'll do this in June when no one is worrying about the cost of winter heating. I wonder how bad it will be in the fall."

I soon found out. Having used fuel oil to heat my home for nearly forty years—and knowing that my parents had used it for decades before that while I was growing up—I knew that fuel oil had always been one of the cheapest products to come out of a barrel of crude oil. In the spring of 1976, despite the higher oil prices resulting from the 1973–74 Arab oil embargo, I had paid 18 or 19 cents a gallon for fuel oil, still a relatively cheap source of heat.

Before Congress had lifted all controls on fuel oil prices, it had been assured by experts of the Federal Energy Administration (F.E.A.—the predecessor of the Department of Energy) that abolishing regulations would stimulate competition. The experts said prices might even be driven down; at the worst, they would rise no more than one cent a gallon. Never were experts more wrong.

In the fall, the fuel oil that had gone into my tank for 19 cents a gallon in the spring jumped suddenly to 30 cents. Outraged, I went to see my favorite gas station dealer. Since I had dealt for years with one of the largest independent distributors in our area, I thought I might save a few cents a gallon by switching to direct buying from a major oil company. But I found that the temper of my favorite Citgo entrepreneur was boiling at a higher pitch than mine.

"The sons-of-bitches are just ripping us off, Mr. Cook," he said. "I'm paying the same price you are for fuel oil to heat my station. There's not a god-damned thing you can do about it."

I soon found that he was correct. Many neighbors along our street purchase fuel from different concerns, some with the name of a "major" (major oil company) on the side of the truck. Every time I saw a truck delivering oil, I would stop and ask the driver what his "liquid gold" was costing. The prices, I found, were all the same. Competition? That was a Big Oil–F.E.A. myth. There was no competition.

The second spur to my journalistic instincts came in the bitterly cold winter of 1976–77. Suddenly, in the midst of Arctic temperature, natural gas supplies were cut off for commercial and industrial plants in the Midwest heartland. Factories and businesses had to shut down; some 1.2 million persons were thrown out of work. It was a crisis.

Yet the shortage that had precipitated the crisis was another myth. Newspaper reports—accounts which I later verified—claimed that there was an abundance of natural gas in the Gulf of Mexico but that wells there were being capped as producers indulged in a force play for higher prices.

Such was the situation when President Jimmy Carter visited the stricken Midwest region. I recalled how the president in his first fireside chat had worn a heavy sweater; how he had painted a bleak picture of our energy resources, and how he had urged everyone to set back the thermostat to 65 degrees. He had appealed to all Americans to conserve and to devote themselves to "joint efforts and mutual sacrifice."

It was natural, therefore, that he should be asked during his Midwest visit about the natural gas shortage and reports that supplies were being deliberately withheld. I watched him as he replied on television, with no visible trace of concern, "Well, you can't blame a businessman for wanting to get a higher price."

Not even if such pursuit of higher private profit throws

1.2 million persons out of work? Was no line to be drawn between greater profit and the public welfare? And just who were the Americans who were supposed to devote themselves to "joint efforts and mutual sacrifice"?

These, it seemed to me, were logical questions, but no one was asking them—not on TV, not in the newspapers. The nonchalant acceptance of the status quo on virtually every level of the media frustrated me, and I decided it was time to take a jab at the Jimmy Carter who had run as a populist and was beginning to act like Herbert Hoover. So I wrote a satirical article and sent it to the Op-Ed page of *The New York Times*. The *Times*, after a long delay, informed me they wanted no part of it.

My frustration increasing, I sent the piece to *The Nation*, a liberal weekly magazine for which I had written for years simply because it would ask questions and probe into issues that the larger media ignored. Again I was stymied. The gremlins in the Post Office must have devoured my article; it simply disappeared into the great unknown, and by the time I learned of this vanishing act, the crisis had passed. Natural gas prices had been raised, the gas had begun to flow, businesses had reopened—and my satire about a crisis that had passed was outdated.

I was disgusted enough to give up, but fuel oil prices continued to gallop upward at a pace that was literally a national disgrace. It bugged me so that I took another tack and jabbed feebly at the oil monolith. It was the merest pin prick at first, but from that pin prick there grew a series of articles that brought me in 1980 the New York Newspaper Guild's Page One Award for crusading journalism. What follows is the story of how this came to happen and of what I learned in the process about the nation's energy problems.

The First Look

2

I might never have begun to look seriously into the multibillion-dollar ripoff of the American people by the Big Oil consortium had I not made a visit to my mother in the near-zero weather of January 1977. My mother was ninety-three at the time, feeble, badly crippled with arthritis, and she was living alone in the old family home in Point Pleasant, N.J., that my grandfather had started to build in the 1870s. Her wants were few, fortunately, for she was making do on an income of about $3,200 a year, supplemented by some $100 a month in Social Security.

Since she could not get out except to hobble around her yard in warmer weather, I did her shopping and handled the payment of her bills. On this particular day, she had just received a bill from her fuel oil dealer. The price of fuel oil had been taking off like a rocket ever since the 1976 price decontrol, and I knew that excessive heating costs were putting a heavy strain on my mother's limited income. But as I looked at this particular bill, the obviousness of the ripoff hit me with special force.

My mother was being charged 46.9 cents a gallon for No. 2 oil that, less than a year previously, before decontrol, had cost 19 cents. I had been aware, of course, that all of my worst fears about decontrol had been realized in spades; but what caught my eye and triggered my investigative instincts on this occasion was that I had just passed an Exxon station

5

at the corner of my mother's street, only a few hundred yards away. And Exxon was advertising regular gasoline for 56.9 cents a gallon.

I knew that fuel oil was untaxed in New Jersey, but that regular gasoline carried a 12-cent state and federal tax burden. When one deducted the 12-cent tax from the 56.9-cent gasoline price, it was obvious that the return to the oil industry was 44.9 cents for regular gasoline—and 46.9 cents for the fuel oil that had gone into my mother's tank.

Ever since I was a boy, and then ever since I had heated my own home with fuel oil beginning in 1940, I had been aware of one basic fact: No. 2 fuel oil, which requires far less refining than gasoline, had always been one of the cheapest products of the refineries. For some fifty years, to my knowledge, it had cost no more than one-third, at the most one-half, of the price of gasoline at the pump. But now the oil companies were making *more* on my mother's fuel oil than they were on the regular gasoline sold at a station a short walk from her home.

It didn't make sense. The case seemed to me so clear that anyone with the intelligence to add two plus two and get four must see that the homeowner was being victimized by a ripoff of gigantic proportions. So, after I returned home, I tested my conclusions with a veteran fuel oil dealer, a man I had known for thirty years. I asked him whether my deductions were wrong. They weren't, he said. Did he have an explanation? He didn't.

"It seems then that something must be damned wrong somewhere," I remarked, shivering in a forty-mile-an-hour gust and the near-zero temperature.

"You better believe it," my dealer friend said. He added that the refineries were setting the price on fuel oil, that they had kept raising prices steadily and that there was nothing a dealer like himself could do except pass the cost along to the consumer.

There was nothing much I could do about it in that winter of 1977, the Post Office gremlins having disposed of my initial attempt; but, as fuel oil prices continued to rise with no end in sight, I became more and more indignant. And in January, 1978, I began to write letters.

I spelled out the dimensions of the fuel oil ripoff to the New Jersey Department of Energy; and on January 24, 1978, Irving Oelbaum, director of operations, attempted to mollify me with some soothing back-stroking. One reason for the price discrepancy, he wrote, was that "there is a price war on in gasoline at the present time which results in less than the normal price for it."

He attempted reassurance. "Our office," he wrote, "is monitoring the price of fuel oil and the price you are paying for it of 49.4 cents a gallon is in line with the market price for it.

In addition the federal Department of Energy is also monitoring the price of fuel oil. . . . They are prepared to take remedial action should the price of fuel oil get out of line. I hope this letter assures you that both the state and federal Governments are aware of the problem and are monitoring it closely.

This sophisticated bureaucratic runaround was to be just the first in a years-long series that would demonstrate the ironclad control that Big Oil exercised over the agencies of government. Though I had yet to discover the extent of that control, I knew that I had been given the bureaucratic shaft. Oelbaum's letter answered nothing, explained nothing. If crude oil supplies were so abundant that there was a price war on for gasoline, why wasn't there a similar price war on for heating oil? And what was this soothing syrup about monitoring prices and being prepared "to take remedial action" if prices got out of line? The figures were clear and undeniable. Anyone with a basic knowledge of grade school arithmetic could see that they were already out of line. The promise of some possible "remedial action" at an indefinite future date was a bureaucratic put-off. It was clear remedial action had already been long needed.

An investigative reporter, as I have been all my life, feels his gorge rise at being brushed off in such fashion. He feels a compulsion to *do* something. Not quite knowing where to turn, I vented my anger and frustration in an article for my local newspaper, the *Asbury Park Sunday Press*, spelling out the dimensions of the ripoff and calling for a federal investigation. The *Press*, trying to present the other side of the story, ran a reply from Edmund W. Renner, executive vice president of the Fuel Merchants Association of New Jersey.

Renner's immediate knee-jerk response was that I was an interfering ignoramus who knew nothing "of the petroleum industry and government regulations over it." There were, according to Renner, two facets of the problem of which someone less ignorant than I would have been aware: Imported Arab oil was much more costly than it had been previously—and then there was that big, bad government with its hampering regulations that kept ceiling prices on gasoline.

It was true that I had no special expertise in the energy field, but common sense told me that Renner was blowing air out of his ears. Since fuel oil and gasoline come out of the same barrel of crude oil, what did higher Arab prices have to do with making fuel oil more expensive than gasoline? And what did federal ceiling prices on gasoline have to do with the issue when there was a gasoline price war on and gas stations couldn't charge even the permitted ceiling prices?

Ignoring such obvious questions, Renner contended that the refusal of the bad old federal watchdog to permit prices on gasoline to "tilt" upward during the summer of 1977 had forced "major producers . . . to include certain production costs of both gasoline and fuel oil in fuel oil prices because of the language of the control regulation." Renner, like

Oelbaum, explained that the federal Department of Energy
(D.O.E.) was using a "price index monitoring system" to
check on fuel oil prices; but then, in a fatal slip, he added:
"Under this system, the actual selling price is too high."

I leaped on that admission in a subsequent article; but
Renner, when he saw it in print, almost had cardiac arrest.
He protested to the *Press* that some qualifying phrases had
been inadvertently left out of that damning sentence and
that he had not meant, Heavens no, to admit that fuel oil
prices were "too high."

My jousting with the oil companies brought some unexpected dividends. It seemed that a number of readers were
as incensed as I was. John R. Laird, of Barnegat, N.J., raised
a significant point. Retired, he was living in a mobile home
heated by keroscne, a fuel that traditionally had sold for a
couple of cents more a gallon than fuel oil, but that had
always been, like fuel oil, one of the cheapest products of the
refineries. Laird wrote, "I own a mobile home and live in it
the year round, and the last delivery I had . . . was 53 cents
a gallon. . . . This is only 2 cents a gallon cheaper than
gasoline and no tax. Most of us who live in mobile homes
are on Social Security and fixed incomes. I hope somebody
will look into this ripoff."

Since thousands of retired persons live in the central
New Jersey area, many of them in mobile homes like Laird's,
I wrote the congressman from our district, James J. Howard,
sending him a file of my letters and the printed exchanges
I had had on the issue. I added:

To me, the whole situation spells out one of the cruelest ripoffs
of the American consumer perpetrated by the major oil companies. People simply have to have fuel to heat their homes in
winter—and they are helpless to combat a system that, in
essence, makes fuel oil more expensive than gasoline. Since this
is a situation that affects thousands of your constituents, I am
soliciting your good offices to build a fire under the federal
Department of Energy.

I will give Representative Howard credit: He tried. But not even a congressman could galvanize the sluggish Department of Energy—already so infiltrated it was practically owned by Big Oil interests—into taking any kind of "remedial action." It was six long weeks before Representative Howard even got an answer out of D.O.E. By this time, of course, the winter heating season was over; and, furthermore, the answer, dated May 11, 1978, virtually duplicated the unresponsive rationalizations that Oelbaum had given me. Douglas G. Robinson, assistant administrator of the Economic Regulatory Administration of D.O.E., began by citing this background:

On July 1, 1976, the Federal Energy Administration [the predecessor of D.O.E.] determined that supplies of middle distillate fuels, including heating oil and kerosene, would be adequate to meet the nation's requirements and that competition would be adequate to protect consumers if price and allocation controls were removed.

Robinson neglected to point out that F.E.A.'s faith in the free-enterprise system had led it to predict that competition might drive prices down; at the worst, they would rise by no more than a cent a gallon. He ignored the fact that prices had zoomed from 19 cents to around 50, an escalation that made F.E.A.'s faith in "competition" and stable prices look ridiculous.

However, Robinson assured Representative Howard of D.O.E.'s "concern" over fuel oil prices ("concern" is a cheap political coin); but then he parroted the Big Oil companies' alibi, writing, "The prime reason for this [rise in prices] has been the increase in the world price of crude oil, which is controlled by the OPEC cartel." (Those intractable Arabs again; it was all their fault.) Robinson continued, "There are several reasons, *other than that a disproportionate amount of costs are being passed through on heating oil,* why No. 2 heating oil is more expensive than gasoline without taxes. One reason is that the supply of gasoline is currently abundant and most suppliers are not charging maximum ceiling prices." (Italics added.)

This admission that "a disproportionate amount of costs" was being loaded on to home heating oil seemed to me like an official plea of guilty. Why, if supplies were so plentiful that gasoline couldn't even sell at ceiling prices, should home heating oil be priced sky-high? The inevitable answer, it seemed to me, was that the oil companies were out to maximize profits regardless of human costs. They could not charge what they wanted for gasoline because there was a glut and a driver could shop from station to station in a price war. But the homeowner cannot move his house to seek the best bargain. If the refineries could uniformly fix the price of heating oil, as all the evidence indicated they were doing, then the homeowner was an imprisoned victim of an unconscionable system.

None of this logic was reflected in Robinson's reasoning. He continued to stress the obvious. Gasoline and fuel oil were seasonal products, he wrote; more gasoline was needed for summer driving, more heating oil for winter heating. There were added distribution and transportation costs associated with fuel oil. Robinson ignored the record of fifty years' experience: these things had always been so and yet fuel oil prices had never in those fifty years even remotely approached the price of gasoline.

Robinson concluded his letter to Representative Howard by playing on his tuba the old "monitoring" refrain. "We intend to continue monitoring this situation and will take whatever action is necessary, including the reimposition of controls, if middle distillate prices begin to rise at a rate not consistent with increases in proportionate refining costs."

It was enough to make a man's head whirl. Having already admitted that "disproportionate" costs were being loaded on to home heating oil, Robinson wound up with that meaningless pledge of action *if* distillate prices were not "consistent" with the rise in refining costs. The first admission seemed to rob the word *consistent* of any meaning.

This exercise in bureaucratic double-talk did not please Representative Howard any more than it did me. In sending me a copy of Robinson's letter, Howard noted

I was particularly interested in the mention of the reimposition of controls on middle distillate prices but disheartened by the lack of mention about refining cost controls. Considering the turn of words in this particular sentence one could assume that if the price of both continues to rise nothing is triggered to help consumers.

The congressman was perceptive. The fuel oil gouge, bad as it was, had barely begun. Heating oil prices continued to run wild, completely out of control, defying justification. In the fall of 1981, as this is being written, heating oil is priced at $1.23 a gallon in New Jersey ($1.28 in New York City) at a time when I can buy regular gasoline, still bearing its 12-cent New Jersey tax burden, for $1.21. Considering the tax equation, that means a 14-cent-a-gallon ripoff right at the start, regardless of what other "disproportionate" costs the refineries choose to load on heating oil. As Representative Howard had written, "Nothing is triggered to help consumers."

Mum's the Word

3

Having appealed to the state Department of Energy, having written to my congressman and called local attention to the ripoff issue, I figured that I had done about all one man could do. I had been given the classic bureaucratic runaround that answered nothing and made only the most meaningless promises. In such cases, most Americans decide it is useless to continue fighting City Hall, throw up their hands and say, "Oh, to hell with it."

I probably would have, too, had I not in the early summer of 1978 had lunch with Victor S. Navasky, the new editor of *The Nation*. Navasky was looking for article ideas; when I told him about my experience with Big Oil and D.O.E., he said, "Well, why don't you write a piece for *The Nation*?" So I did—and the article started an entirely unexpected chain of events.

Published in *The Nation*'s July 22–28, 1978, issue under the title "Fuel Oil: The Rip-Off Runaround," the article aroused the fury of Joel R. Jacobson, New Jersey commissioner of energy. Jacobson was angry with me because I had put his deputy, Irving Oelbaum, in a bad light. Jacobson informed me that Oelbaum was a very valuable and trusted aide of his, and he thought I had done him a great injustice.

I answered Jacobson in equally angry terms. I spelled out the situation for him, I explained that I had not received a single definite answer to any point I had raised either from his department or the federal D.O.E., and I wound up by asking what the devil he could expect me to call this except a runaround. There was a long silence. Then, late in 1978, Jacobson wrote me a surprising letter. He said that it was sometimes possible to stir the bureaucratic behemoth into action and that, as a result of my article and letter, he was going to hold three days of hearings at which representatives of all the major oil companies would be asked to testify and justify their prices.

It was the first time any state commissioner in the nation had taken such action, and it seemed to me that, with Jacobson's official clout, it might be possible to get some answers. This was, as it turned out, a delusion of the naive: What the hearings demonstrated was that Big Oil could thumb its nose at official authority with impunity.

Jacobson held his first hearing in Trenton, the last two at Seton Hall University in Newark. I sat and listened as spokesmen for the major oil companies took the witness stand. There they were, clad in their conservative, executive-status suits, their lawyers by their side. They all spoke forcefully, with great conviction. They knew to the last human digit the number of their employees in the state of New Jersey, the number of their refineries, their capacity in millions of gallons. They all inveighed against the restrictions placed on Exxon, Shell, Mobil, Texaco, et al., by a federal government that insisted on keeping ceiling prices on gasoline. It was the dirty hand of government regulation, they charged, that kept the beautiful free-enterprise system from operating in its infallible way for the good of us all. That was why we had fuel problems.

But once they had gone through such oratorical flourishes, these powerful executives of some of the largest multinational corporations in the world became flustered and far less certain when they were asked a simple question. The question, repeated over and over to various witnesses, was, "Can you tell us what it costs to produce a gallon of No. 2 heating oil?"

Then, suddenly, these experts who had answers for everything ran out of answers. Time and again, Commissioner Jacobson and his aides heard the same refrains: "I'm sorry, sir, but I don't have those figures," or, "I'm sorry, sir, but we just can't break costs down that way."

Joseph Hinton, general manager of planning and finance for Mobil, put the industry argument this way.

Large refineries like ours at Paulsboro contain many operating units. While we can determine the cost of operating a given processing unit, *there is no way to determine the cost of a single refined product*. . . . The actual price of a given product is determined by competition and government regulation. [Italics added.]

H. Scott Ingersoll, manager of oil products and sales services for Shell, put the no-figure alibi this way.

So you are making from a barrel of crude a wide spectrum of products and our ability to slice that barrel up and say precisely what costs were allocated to making gasoline and what costs were allocated to making the No. 2 oil, and what costs were allocated for making lubricants, we don't have that ability.

As the questioning went on, a new genie popped into this Arabian Nights tale. Industry spokesmen explained that they computed costs on a "volumetric" basis. That is, they knew what crude oil cost and what it cost to run the refinery; they knew how many gallons of crude were processed and how many gallons of total product were produced. But since they couldn't tell what any individual item had cost, they could price the final products virtually any way they pleased, assigning costs at will to unregulated home heating oil.

Significantly, the industry spokesmen asserted that the Department of Energy, supposedly the federal watchdog, had approved this "volumetric" way of computing costs. Listening to this, I wondered how D.O.E. had had the gall to tell Congressman Howard it was monitoring fuel oil prices. How could it monitor anything under this "volumetric" system that did not break down and allocate costs to any specific product?

Commissioner Jacobson and his aides attempted to pierce this volumetric smoke screen. A couple of exchanges will illustrate the way it went. Commissioner Jacobson was questioning Robert Hulting, staff director of profit planning and pricing for the eastern regional office of Amoco. Jacobson said his office had tracked the reseller tank car price of both gasoline and heating oil since heating oil had been decontrolled on June 30, 1976. He had found that, while gasoline had risen in price 6.3 cents a gallon, heating oil had jumped 9 cents.

"My question is, if you assign costs volumetrically, why should there be a disparity in these two prices?" Jacobson asked.

Hulting tried to avoid the implications of this analysis by suggesting that, because there is a greater demand for heating oil in winter but virtually no demand in June, this would affect the price and possibly explain the differential.

"I heard your earlier explanation about letting the free market apply," Jacobson said, "and I understand that, and I do think, though, it's possible for a cynical mind to reach the conclusion looking at these two figures that perhaps controls do protect the consumer more than decontrols. Would that not be a logical response?"

Hulting had to concede that perhaps one could look at it that way.

Oelbaum pointed out to Hulting that heating oil prices had jumped 1 cent a gallon for three consecutive months— August, September and October 1978, long before the heating season really began, long before heavy increased demands for heating oil might have been expected to affect prices. He emphasized that this had happened despite the fact that there had been no increase in crude oil prices by OPEC, despite the fact there had been no comparable increase in gasoline prices.

Hulting kept insisting that he didn't think heating oil had increased in price disproportionately compared to gasoline if you looked at the long-term picture and that, anyway, "the free marketplace" would bring prices down if they got out of hand. He had never seen the free market not operate that way, he said.

It was obvious that none of this made much sense to Jacobson and his assistants. Jacobson made his disenchantment clear to R. C. Knowles, regional manager of Exxon, when he said, "When we continue to ask questions about the price, frankly, we get nebulous answers, and no matter how we push, we can't get the answers. If I were to ask you what were the production profits per barrel, I suspect that you wouldn't be able to tell me, would you?"

"Well, that's not my field. I don't really have that figure," Knowles said.

The contrast between the executive-suite corporation types and the spokesmen for the working stiffs was so sharp as to be startling. Jerry Ferrara, executive director of the New Jersey Gasoline Retailers Association, was blunt and forthright in contrast to the volumetric dodgers.

Our observation is that the suppliers [the Big Oil spokesmen] have been less than candid in their replies. . . . Nothing that I have seen or heard indicates any crude oil shortage through the 1980s. . . . The wholesale price of gasoline has risen 140 percent from 1973 to now, excluding taxes. . . . Our association has challenged the rise in prices, but prices go up and up. There has been possibly $1.5 billion of overpricing.

Arthur Cole, first vice president of the New Jersey Industrial Union Council (A.F.L.–C.I.O.), representing two hundred thousand workers in the state, struck hard at the manner in which refineries had loaded costs on deregulated home heating oil, victimizing millions of home owners. He quoted directly from the Federal Energy Agency's announcement when heating oil was decontrolled. He gave special emphasis to this F.E.A. statement to Congress: "If all of F.E.A.'s findings about projected supplies, demand and competitive forces are wrong, price might rise by as much as *1 cent a gallon.*" (Italics added.)

As Cole pointed out, the continued rise of heating oil prices had made F.E.A.'s worst-case scenario look positively ridiculous. "In the year following decontrol," Cole testified, "the reseller tacked-on price rose almost 5 cents a gallon in Newark. Today the average retail price of home heating oil has gone up over 12 cents a gallon in New Jersey or twelve times the F.E.A. projected price in a short two and a half year span. There is not an oil-burning home in this state which is escaping the effects of these price increases."

The testimony made it clear that free-enterprise price competition was a myth. Those 1-cent-a-month price increases in the fall of 1978 had been uniform across the board, no matter what refinery produced the oil; and the prices had not stopped—they were still rising in perfect cadence.

Much independent checking of my own verified the disclosures in Jacobson's hearings. When my fuel oil jumped to 52.4 cents a gallon in January 1979, my dealer told me, "Before the Arab oil embargo [in 1973–74], you used to be able to shop around. You might be able to get fuel oil from one refinery 3 cents a gallon cheaper than from another. But not any more. They all set the same price. If one raises the price 1 cent a gallon today, they'll all have the same figure tomorrow morning."

There was then—and has been ever since—a great attempt by the oil companies to explain high fuel oil prices by passing the buck to the deliverer. The alibi goes that he has to pay increased costs for his tank truck, his drivers; that he has to pay high interest rates to finance his fuel purchases until he can get reimbursed by his customers. Most of this is pure hogwash.

Even as Jacobson's hearings were being held, I got into a

discussion at a small luncheonette with a fuel oil dealer whom I've known for years. I expressed my horror at having to pay 52.4 cents a gallon for fuel oil.

"Do you know what the refineries are charging *me*?" the fuel oil dealer said, jabbing a finger at his own chest. "Forty-six point nine cents a gallon! I have to sell it at fifty-two point four. That's only a five-and-a-half-cent margin. Now in the winter, in cold weather, I'll sell maybe seven thousand to nine thousand gallons a day. But I have to pay the driver, the insurance on the truck, the repairs on the truck, the gasoline to drive the truck—and when I'm making deliveries like that, the gasoline alone comes to twenty-five dollars a day. How can I do it on a five-and-a-half-cent-a-gallon margin? These jerks are going to bring us all down."

He was just warming up. "I'll tell you something else," he said. "I used to be given thirty days to pay for my oil. Not any more. Now it's ten days. It used to be that if you paid in ten days you got a discount—sometimes a double discount—but today you get nothing. When I scream about it, my supplier says, 'Well, go to the bank and borrow the money.' What, at these rates? Fifteen to twenty percent? That would put me out of business. So I have to try to get cash on delivery from my customers. I've had some of them for twenty years, and it's pretty hard to explain to them why I can't give them credit any more."

At this point, the telephone in the luncheonette rang. The call was for the fuel oil dealer. He answered it and came back shaking his head. "That was a customer of mine," he said. "He already owes me a hundred dollars for fuel oil—and now he needs another hundred gallons. That's fifty-three dollars. I know he's good for it, but he can't pay me now. He has a seven-week-old baby in the house. So what am I going to do?"

He threw up his hands helplessly.

To return to the Jacobson hearings, there were a couple of threads that ran through the testimony to which not too much attention was paid at the time. My own attention—and that of other reporters who covered the hearings—was concentrated on the heating oil pricing situation, the main focus of the probe. None of us could read the future then, none of us had any idea that 1979 was going to be a year of "crisis" with gasoline "shortages" and long lines of angry drivers waiting at gas station pumps. As a result, there were niggling little bits of testimony that did not seem vital at the time but would assume far more significance later.

One was the admission by oil company witnesses that there was so much gasoline on hand that stations couldn't even charge ceiling prices for it. There was a consensus that, except for a temporary shortage in unleaded gasoline, supplies were so ample that the market was literally saturated. In fact, as I drove into Newark for the hearings, regular gasoline was selling for 59.9 cents a gallon (12-cent tax included) at a time when I was paying 52.4 cents for some heating oil—a disparity even more glaring than the one that had caught my attention in 1977. But there was a hidden gimmick in this gasoline price that I did not understand.

A hint was contained in the testimony of Exxon's Knowles. He said at one point, "As of June [1978], the industry had about $1.5 billion of banked pass-through that they could have passed through to the customer that has not been passed through because no one, I guess, feels the marketplace could absorb it. So the price of gasoline could go up even more, you know, if it was felt by any individual company that the marketplace could stand it. . . . So, under the controls of cost pass-through, there's about a billion and a half dollars as of June in the bank, so to speak, that could be passed through by the industry."

I plead guilty to not understanding the significance of the point at the time. In exculpation, I can only ask: How could any sane man, even a suspicious one, conceive that our vaunted free-enterprise society would sanction a scam that would permit Big Oil to "bank" in its accounts what it couldn't charge now so that it could sock it to the public at an opportune time later? Yet this was precisely what Knowles was referring to. A beneficent D.O.E. regulation permitted gas companies not able to get the ceiling price for gasoline to save up the difference between their selling price and the ceiling price so that it could be added to prices later. All that would be needed to raise prices to the ceiling—and then drive them still higher by tacking on these "banked" balances—would be a "crisis," a shortage so severe that drivers would be clamoring for gasoline at whatever price.

It was as if a department store had held a 20-percent-off sale, as many were doing in the pre-Christmas season of 1981, because customers wouldn't pay the regular price. And it was as if, a month later, the department store had tacked on the extra 20 percent above its regular prices to recoup from the sale. Ridiculous? Of course. It couldn't be done because department stores compete in the marketplace as oil companies do not.

The whole concept of "banked pass-through" in gasoline accounts was so foreign to my experience that I didn't understand its importance at the time. And, if I had understood it, it wouldn't have seemed to matter greatly because all sources agreed there was such a glut of gasoline that it seemed unlikely any use could ever be made of those "banked" credits. Amoco's Hulting had summed up the outlook of the industry when he said, "Anybody can get crude oil today. It costs money, but it's available. I'm not saying we don't have an energy problem long-term, because we do, but it's not a problem today of getting crude oil."*

Despite such testimony, the big oil companies at this precise time were launching a power play that saw them all dancing in cadence like Rockettes in a chorus line. Shell Oil

* *Priceline,* the McGraw-Hill newsletter that forecasts trends, had reported in February 1978 that the global oil glut would last for at least two more years. It added that supplies looked adequate through the 1980s and that proven oil reserves could last close to forty years.

started it. It announced that, as a result of unusually mild fall weather in 1978, American drivers had been consuming a lot more gasoline than was normal. A shortage had developed in unleaded gasoline, and so, Shell said, it was going to have to ration gasoline supplies to all its service stations. Other members of the Big Oil Rockettes hastened to dance to Shell's tune.

I happened to walk into my favorite Citgo station a couple of days after the Shell announcement, and I heard the proprietor screaming over the telephone to his supplier, "How the hell am I going to do business? You can't cut me back like that. I have to supply all the gasoline for two towns here. The police cars have to have gas. And what about my customers? You've *got* to do something!" He slammed down the phone in disgust. "Jesus Christ!" he said to me. "If they keep up this rationing bit, you're going to have lines waiting at the gas pumps again."

A day or so later I walked into a station that was being supplied by Mobil. The owner was furious. "They tell me I can have only eighty percent of the quota I used in June," he said. "But I'd just opened the station in June—and so we sold only eighty thousand gallons. Last month we sold a hundred seventy-one thousand. How am I going to take care of my customers if they give me only seventy thousand gallons a month?"

The Big Oil power play had its effect. The Carter administration either panicked or fell willingly into line with the Big Oil Rockettes. Alfred Kahn, Carter's inflation czar, told Congress that price controls on gasoline would have to be lifted to ensure the nation of adequate supplies. The White House announced that President Carter intended to ask Congress at its next session to lift all controls. The administration estimated that such action would result in picayune increases—no more than 2 to 4 cents a gallon. Those who recalled F.E.A.'s 1-cent estimate on fuel oil weren't reassured.

Congress had its own reasons for being skeptical. A staff report prepared for Rep. John D. Dingell (D., Mich.), chairman of the House subcommittee on energy and power, had charged in December that "major" oil companies (unnamed) had overcharged American consumers $2 billion in the last four years. It also alleged that some officials of the Department of Energy had condoned practices that chiseled American consumers out of some $2 million a day.

Against this background, it seemed amazing that the media began to be flooded with estimates that gasoline should really be selling at $1 a gallon and heating oil at 60 cents plus. The Carter administration seemed to view such inflationary estimates with equanimity. It adopted, with hardly an effort at disguise, the philosophy that the more Americans were hurt by higher prices, the more they would be forced to conserve and the sooner the crude oil supply problems would be brought under control.

It was a policy either of gross stupidity or gross lack of conscience. Eliot Janeway, the noted economist, made an acerbic attack on Alfred Kahn for his abject surrender to the Big Oil ploy. Janeway said, "You can't do anything about inflation if the price of oil is allowed to go up because the cost of oil affects the cost of everything." He characterized the oil companies' argument that they had to be free of regulations to be encouraged to drill for more oil as so much "baloney." To talk about fighting inflation at the same time that you are lifting ceiling prices on oil is like "putting on your skates to go jogging," he said.

Exxon News illustrated Janeway's point perfectly, if unintentionally. In its report to stockholders in December, it noted, "United States agriculture is the most energy-intensive in the world. From farm to ultimate consumer, all of its activities account for about 15 percent of total U.S. energy consumption. Oil and gas combined meet about 80 percent of agriculture's energy needs." This translated into a farm consumption of 12 billion gallons of oil a year.

It did not take much imagination to envision what would happen to food prices when those 12 billion gallons of oil used on the farms should escalate to a price of a dollar or more a gallon. The American housewife was already screaming about exorbitant supermarket prices, and many elderly persons living on small pensions and Social Security were already buying and eating canned cat and dog food. It seemed obvious that the script being prepared for us could bring only more tragedy, worse disaster.

Only one dissident official voice was raised in the midst of this insanity. Brock Adams, secretary of transportation in the Carter administration, appeared on NBC's morning television show "Today" and took two industry giants to task. He swatted at the motor moguls of Detroit with one hand and at the big oil companies with the other. Adams wanted Detroit to produce cars that would give the American motorist fifty miles to a gallon of gasoline. He insisted this goal was not unreasonable because there were already in existence three or four new-style motors that would give greatly increased mileage; but, he charged, the closely interlocked motor baronies of Detroit refused to undertake any radical revamping of their present engines.

As for petroleum, Adams said, the international consortium composed of a few major oil companies exercises control over the market. They have the power to set prices and determine the allocation of products. Adams added that we had the knowledge to make synthetic gasoline and the ability to stretch present gasoline supplies by adding alcohol to the mixture. But the nation cannot utilize these alternatives, he said, because the multinationals control the market; and, if a competing product should threaten their control, they'd slash their prices enough to drive its manufacturers out of business.

I wondered in a *Nation* article at the time how Adams had ever happened to become a member of Carter's cabinet and how long he would last. The second question was quickly answered; he was soon gone. It was a pattern that I was to find repeated time and time again. Any critic of the Big Oil–Carter script who had the temerity to speak out soon had his official hat handed to him and was banished to the boondocks.

As for this first 1979 gasoline shortage scare, it quickly faded and was all but forgotten by the time the "real" 1979 "crisis" hit the country a few months later. But for now, supplies were just too abundant; the ploy was just too obvious. And so the Big Oil Rockettes had to beat a retreat into the wings for the moment. But only for the moment.

There was still that $1½ billion of "banked" credits to collect; there were still the additional billions to be made by driving gasoline and heating oil prices into the stratosphere. Big Oil hadn't given up. It was merely waiting.

The Oil Stonewall

4

In early January 1979, Joel Jacobson was an extremely frustrated man. When he had failed to get responsive anwers from oil company spokesmen at his December hearings, he had threatened to subpoena the records of the recalcitrant companies, only to find that his hands were tied.

American law in its infinite wisdom has sanctioned a most comprehensive business blackout through a concept known as *proprietary information.* What this means is that a company does not have to disclose the details of its business to anyone, on the theory that, if it did, some competitor might discover its secrets and take unfair advantage of it.

Such a protective shield may be justified in cases where a company is protecting a process known only to itself, but in the case of *Jacobson v. Big Oil,* no secret processes were being protected. The legal stonewall was simply allowing the oil companies to hide the manner in which they were amassing huge profits at the expense of the inflation-riddled American economy and the average American homeowner.

Balked, Jacobson had taken his case to the House subcommittee on energy and power shortly before Christmas 1978. His hearings had convinced him that heating oil prices were accelerating at a pace that could not possibly be justified by costs, and so he called for a reimposition of controls.

"It is sheer folly to permit a national policy on home heating oil—a policy which says, in effect, 'the sky's the limit'—to stand as we watch it erode the livelihood of the American consumer," Jacobson said.

He was to testify to the same effect time and time again in the years ahead. Congressional committees heard him out; many congressmen agreed with him—but there it ended. There was never any action. The stonewall in the executive branch of the Carter administration was complete.

When I talked to Jacobson in early January 1979, after his first appearance in Washington, he discussed the overall petroleum situation and made some perceptive predictions. Neither of us knew then that the spring of 1979 would find the nation suddenly submerged in an oil "crisis" that would have long lines of frustrated motorists snarled up at the gasoline pumps, but Jacobson, even this early, sensed what might be coming.

Everybody keeps telling us there is a shortage, but the information I get is that there is no real shortage. It's a contrived situation to drive prices up. For instance, say a refinery is getting oil from Saudi Arabia at the OPEC price of $13.35 a barrel. The spot market price is $18 to $20 a barrel. [Actually, at this time, it had gone as high as $23.] So the refinery uses the spot market price as the justification for raising prices of its refined products, giving it a huge inventory profit.

You saw where the profits of the big oil companies were up 30 percent in the last quarter of last year. Well, wait until the reports for the first quarter of this year come out. You haven't seen anything yet.

Actually, Jacobson's reference to oil company profits was extremely conservative. As more and more reports were made public, all showing an enormous jump in fourth-quarter profits, a shock wave ran across the country, and President Carter was confronted with the situation at his January 26 press conference. Question sixteen concerned those fourth-quarter oil profits.

Mr. President, the fourth-quarter profits are out for the big oil companies, and I won't mention names but they reached 48 percent, 72 percent, 44 percent, 134 percent. Given the fact you've asked the country to make sacrifices, prepare for lean and austere years, I was wondering if you'd give us your reaction to those profits—profits that size when American workers are being asked to hold wage increases to 7 percent?

Jimmy Carter waffled on this one. "Well," he said, "in the new energy bill we've obviously had some basis on which to increase incentives for production of oil and natural gas. . . . And above a certain point in earnings, of course, the income tax levies against even the oil companies are partially adequate."

He "hoped" for "a more stable production rate" and for the ability "to keep all industrial profits at a reasonable level with none being exorbitant." He couldn't anticipate whether increased production and "more competition" would lower prices. He ended his answer by adding piously, "But I would, obviously as all Americans would, like to see a good balance between prices and profits."

In the circumlocutions of this answer there was no acknowledgement that the enormous profit increases cited to him in the question indicated that oil company profits were already exorbitant. There wasn't a hint of concern about excessive Big Oil profits, nor a hint of concern about the welfare of the average American who was being asked to limit his wage demands to 7 percent. There was only that sickening piety at the end.

There had been occasions on which Jimmy Carter had sounded far more strident. In a news conference on October 13, 1977, he had lashed out at the big oil companies, accusing them of attempting to stage "the biggest ripoff in history." He had warned of "potential war profiteering" and had added, "The oil companies apparently want it all." He had sounded like the Jimmy Carter who had campaigned in his Halloween suit as a populist, but it was, as Shakespeare would have said, all "sound and fury, signifying nothing."

Carter's actions had not kept pace with Carter's fighting

words, and anyone who seeks an answer to this conundrum may find it in the little-recognized ties of Jimmy Carter and most of his administration to Rockefeller–Big Business interests, heavily saturated with oil. One link to which not much attention was paid at the time was to the mysterious Trilateral Commission, about which the public knew virtually nothing.

In 1972, David Rockefeller, head of the powerful Chase Manhattan Bank, had convened a meeting of leaders from North America, Western Europe and Japan to analyze world problems, including energy. The Trilateral Commission was formed a year later, with Zbigniew Brzezinski its first director. Rockefeller himself was chairman of the North American section.

The commission brought together 275 prominent businessmen, labor leaders, scholars, statesmen and politicians. It had a three-year budget of $1.67 million. Among the largest contributors, foundations contributed $644,000; corporations, $530,000; individuals, $220,000; and $84,000 came from investment income. The Rockefeller interests led all contributors. The Rockefeller Brothers Fund donated $180,000; the Rockefeller Foundation, $100,000; and David Rockefeller personally, $150,000.

Politician-members of the Trilateral Commission severed their connections to the group when they assumed public office, but many returned to it once they again became private citizens. The fraternization implicit in such a set-up could hardly fail to influence the thinking and actions of ex-Trilateralists in office; and so it seems significant that nearly twenty of this private brotherhood served in the top levels of the Carter administration.

Carter himself had belonged to the commission. So had his vice president, Walter F. (Fritz) Mondale, who in 1981 was to demonstrate his affection for Big Oil by serving as a paid consultant for interests trying to build the Alaskan gas pipeline at consumers' expense. Brzezinski, the first Trilateral director, became Carter's national security advisor. Cyrus Vance, another of the brotherhood, served as secretary of state. Harold Brown, yet another of the clan, became secretary of defense. If one considers personal ties and influence important in the affairs of state, this virtual take-over of the entire top echelon of the Carter administration must seem significant indeed. But this was not all.

In what might seem like an influence-wielding pincer movement, a second enfolding wing operated through the Chase Manhattan Bank and the Rockefeller family ties to the oil industry. Daniel Guttman, a Washington attorney who has represented state and local agencies in energy matters and who later served as chief counsel for a Senate investigating committee, traced the interlocking power structure of Big Oil in testimony before a Senate committee in 1979.

Guttman cited a 1978 report by the Senate's Metcalf subcommittee showing that executives of Exxon and Standard Oil of Indiana served on the board of directors of Rockefeller's Chase Manhattan Bank. The Rockefeller Family Group, the largest single voting interest in Chase, was also among the top voting interests in Exxon, Mobil and Standard Oil of Indiana (Amoco). Chase Manhattan itself was among the top voting interests in Exxon, Mobil, Standard Oil of California, Shell, Standard Oil of Indiana and Texaco. In other words, the most powerful oil companies in the nation formed an interlocking combine that virtually assured unified control of the nation's most important energy sources.

It is only through an understanding of this cartel structure of Big Oil, added to the potential influence exerted at the top by the Trilateral Commission's fraternal ties, that one can begin to make sense of the manner in which crude oil supplies were manipulated and multibillion-dollar ripoffs were allowed to run unchecked—indeed, were protected— during the years of the Carter presidency.

Guttman traced the beginnings of this influence back to the days of the Ford presidency. Ford's Federal Energy

Administration, the body that had whisked its wand and decontrolled heating oil, had needed facts and figures about the oil industry in 1974. So it had turned to an outside consulting firm to furnish the facts on which it could base policy. The firm chosen was one the American public had probably never heard of—R. Shriver Associates (RSA), of Parsippany, N.J. How had RSA popped out of nowhere to assume "a most crucial intelligence-gathering function," Guttman asked.

RSA was selected because the needed oil industry information was available only from the Chase Manhattan Bank data file. If the justification is to be believed, the executive branch chose to abdicate its intelligence over untold billions of dollars of policy-making. . . . As the justification [furnished by Carter's D.O.E. from earlier records furnished to Senator James G. Abourezk] stated

> The Chase Energy Information System has been developed over the past 20 years, and is a unique, comprehensive, detailed and reliable source of energy information. To reconstruct the valuable and required tool would require over $1 million in investment in addition to the time required.
> Because of their position in the banking world, Chase will only deal with the Government through a third party, and has so designated R. Shriver Associates.

In other words, Chase Manhattan, involved up to its multibillion-dollar neck in the oil business, virtually dictated to the government the manner in which the government was to get the information it wanted.

As a result of Chase's insistence on the employment of the Shriver firm, one of Shriver's employees, a man named John Iannone, became an important figure on the Washington energy scene. Although he was an outside consultant, Iannone was assigned office space at the Federal Energy Administration and became in appearance, if not in actual fact, a member of the government's official energy establishment. His status did not change when the Carter Administration took office. From April 1974 to July 1976 he worked with Dr. Daniel B. Rathbun, who was deputy assistant administrator for data. Then Dr. Rathbun changed hats— and so did Iannone. Rathbun moved from the federal energy agency to become a vice president of the American Petroleum Institute, and Iannone followed him, becoming a paid employee of A.P.I.

This was a significant change. The American Petroleum Institute is the trade association of the oil industry; it is a registered lobbyist for the industry. It has three regional offices, thirty-four state offices and a budget of $32 million. It is financed by Exxon, Mobil, Texaco, Shell, Phillips Petroleum, Getty Oil, Marathon Oil and others, with Exxon and Mobil its largest financial contributors.

If John Iannone's connection with federal energy policy had been questionable when he wore the cloak of Chase-Shriver, it became virtually scandalous when he went on the payroll of the oil industry's lobbying organization—and still retained his private office in Carter's D.O.E. His cozy relationship with D.O.E. officials continued; he had the run of the offices. Having once been granted the cachet of a member of the establishment, he still carried a semiofficial aura about him and exercised considerable influence on the formulation of Energy Department policies.

Mark Green, of Ralph Nader's Congress Watch, was the first to stumble across the trail of Iannone. He obtained a copy of Iannone's quarterly report to A.P.I., summarizing his activities for the first three months of 1978. In this, Iannone claimed that he had exerted great influence in the formulation of the gasoline pricing structure and other policies. Green's disclosure triggered official probes.

Senator Howard M. Metzenbaum, chairman of the Senate Finance Committee, called on Carter's D.O.E. secretary, James R. Schlesinger, to testify and explain how an A.P.I. agent was exercising such influence over his department. Schlesinger, who had come out of RAND, the Defense

Department's West Coast think-tank, and who had later served in the Nixon-Ford era as secretary of defense and director of the Central Intelligence Agency, had the reputation of being one of the most arrogant of all the arrogant types ever to hit Washington. He demonstrated that now. He brushed off Metzenbaum as if the senator were of no more consequence than a troublesome mosquito. Refusing to appear himself, he sent up to the Hill his deputy, David J. Bardin, head of the Energy Department's Economic Regulatory Administration.

Metzenbaum was furious. He berated Bardin, denounced Schlesinger and blasted the intransigence of the Energy Department. He managed, however, to get Iannone on the stand; he obtained some of Iannone's records; and he charged that Iannone, through his close ties with D.O.E., had effected a change in gasoline pricing that had cost American motorists $600 million annually. Schlesinger's D.O.E., Metzenbaum roared angrily, had "failed and failed miserably" to protect the interests of the American people.

The disclosure of Iannone's activities had a ripple effect in Washington, especially among consumer advocates who had been kept cooling their heels in Schlesinger's outer office while boll weevil Iannone was effecting changes inside. "I'm just stunned," said James Flug, then head of Energy Action. "These people get everything at the earliest possible steps in the regulatory process, having impact on decisions at every step of the way on issues we don't even know exist."

Schlesinger's D.O.E. was bombarded with angry calls for an investigation from Mark Green, Metzenbaum and Rep. John Dingell, chairman of the House subcommittee on energy and power. Even Schlesinger's aide, Bardin, thought there should be an inquiry; and so the inspector general of D.O.E. was ordered to undertake it.

The inspector general had difficulties. Iannone refused to testify, fought a subpoena in federal court and escaped an appearance. The American Petroleum Institute likewise balked at first but finally agreed to release pertinent documents from its files.

Some 160 documents were thus obtained, and they revealed a pattern of infiltration and influence beyond anything that had been imagined. In a report issued April 23, 1979, the inspector general concluded:

- The American Petroleum Institute maintained a regular spy network, titled Federal Agencies Department. Three other agents working for A.P.I.—Henry Rum, Patricia Hammick and Susan Hodges—had infiltrated D.O.E., cultivated contacts there and flashed advance warnings of proposed regulations.
- Iannone himself had been able "to obtain from D.O.E. staff copies of twenty-three draft rulemakings and internal D.O.E. memoranda and studies that had not been made formally available to the public in the last half of 1977 and the first quarter of 1978."
- Iannone had claimed, "Because of the inputs I gave to D.O.E. on middle distillate and Mogas [motor gasoline] monitoring, I was asked to review the final rulemakings for accuracy before they were sent to the D.O.E.'s general counsel." The inspector general found this statement substantiated by A.P.I. documents and by admissions of some D.O.E. officials that they had ordered proposed regulations checked with either Iannone or A.P.I. before they were put into effect.

The picture that emerged was one of a vital government department acting as the lackey of the industry it was supposed to be regulating. The inspector general did not put it so bluntly, of course, but throughout his report he cited admissions by D.O.E. officials that they had consulted with Iannone and A.P.I. Even Bardin admitted that he had directed a subordinate to have the gasoline monitoring formula reviewed by the petroleum institute.

Middle distillate monitoring (MDM) was another ticklish issue. The report acknowledged that there had been "numerous criticisms of the system used" during the 1976–77 winter, and so it was decided to make some changes in MDM regulations. The report continued:

Mr. Iannone claims that he organized an industry working group that reported to a "subcommittee" on MDM. This subcommittee apparently existed within A.P.I.

Mr. Iannone's description of his involvement with MDM is somewhat misleading because of incompleteness. At least two D.O.E. officials, Gerald Emmer and F. Scott Bush, of the Economic Regulatory Administration (ERA), said that they had consulted with Mr. Iannone about the middle distillate formula/ index. Mr. Bush was also an apparent source for Mr. Iannone on MDM-related internal documents found in A.P.I.'s possession.

It seems clear that D.O.E. and Big Oil were in cahoots; and so it is little wonder that neither Representative Howard nor Commissioner Jacobson got any satisfaction when they challenged the department about the cruelest ripoff of all— the continuing, wild escalation of heating oil prices.

The Iannone-A.P.I. intrigue with D.O.E., striking as it was, was only one indication of a much wider evil—the use of consultants who were also oil industry consultants to fashion official policies. Even President Carter's much-ballyhooed National Energy Plan, announced by the White House on April 29, 1977, had its genesis in the brains of oil industry consultants. This was not generally recognized until Daniel Guttman testified in October 1979, and even then the American public was kept in ignorance because the vast majority of the mass media ignored or paid little attention to the disclosure.

"Contractors have been the invisible hands in the preparation of the President's National Energy Plans and Presidential Energy Message," Guttman testified. "The executive branch has neither adequately informed the public of the role performed by contractors in national energy planning, nor of the private interests with which they are affiliated."

Guttman gave specifics. In preparing Carter's first National Energy Plan, a firm called Energy and Environmental Associates received two contracts—one for $194,000, a second for $34,800. D.O.E. admitted it had never inquired into this consulting firm's other associations, but Guttman had discovered by reading the firm's own brochures that its clients included Exxon, Shell Oil and Standard Oil of Ohio.

A second consulting firm, Arthur D. Little, well-known in the profession, had also worked on the president's plan and had received $180,000 for its input. The Little firm, Guttman pointed out, has advised a number of large coal producers and companies interested in nuclear and solar developments.

Energy and Environmental Associates, Guttman continued, was engaged not only to help draft the energy plan but to whip up support for it by assisting "in conducting a series of meetings with four to six corporations. . . ." This assistance also included the briefing of government officials in preparation for these corporate meetings.

E.E.A. remained a favorite of the Energy Department. In 1978, it received another contract that eventually amounted to $1,977,903. "The 'scope of work' for the contract," Guttman testified, "called on E.E.A. not only to run the government, but to evaluate the results of its management."

Guttman added that, when Carter devised his Second Energy Plan, "eight contractors (including E.E.A.) were involved in the preparation." Further contracts were "specifically awarded" to consulting firms to rally support for the president's message.

Guttman's testimony and pressure by Senate and House committees resulted in a study by the General Accounting Office, the congressional watchdog, of the extent to which private consultants were actually running the Energy Department. The comptroller general reported his findings on

November 2, 1979. The report was a shocker—or should have been, had the media paid any attention to it.

For the report stated bluntly that D.O.E., which had some 20,000 employees, had spent 79 percent of its 1978 budget for awards to private contractors to tell it what it was to do and how to do it. The contracts totaled a whopping $8.5 billion. This was the worst record, the G.A.O. stated, compiled by any government agency, and it added:

Each of the five organizations within the Department of Energy appears to be contracting with private firms to carry out some of its basic management functions. These contracts are written so that the contractors are required to perform activities such as program planning and development and establishing goals and priorities. Some of these contracts appear to provide contractors wide latitude for participation in the development of energy policy and offer the potential for allowing the contractor to determine energy policy.

This incestuous relationship between D.O.E. and Big Oil through consulting intermediaries induced the *Washington Post* to run a week-long series of articles in the spring of 1980 exposing the secret government of government by consultants. The series, in turn, sparked Senate hearings on the consultant takeover of D.O.E. Daniel Guttman was chief counsel for the investigation. Sen. David II. Pryor conducted the hearings, and he and Rep. Herbert E. Harris II introduced a consultants' reform bill which, in the nature of things in Washington, died a quiet and inconspicuous death.

In a statement summarizing his committee's findings, Senator Pryor drew a succinct and powerful picture of the dominance of D.O.E. by consultants. He said:

Contractors are running the Department of Energy. They prepare the department's basic plans and budget documents, reports to Congress and congressional testimony. They answer mail from the public and Congress. They are paid to prepare basic components of their own contracts, and contracts for others, and virtually run the contract file room. They prepare the D.O.E. organization charts and the job descriptions of civil servants.

The contractors most heavily relied upon by D.O.E. include those that publicly boast of their oil and utility interest clientele.

In a second statement a month later, Senator Pryor revealed that many of the consultants relied upon by D.O.E. were not only consultants for the oil industry, but consultants for OPEC nations whose pricing policies had thrown the economies of the Western industrial world into turmoil. He spelled out in detail the case of Booz Allen and Hamilton, which had received over fifty D.O.E. contracts "totaling tens of millions of dollars, and involving virtually all of the department's areas of work—fossil energy, solar energy, nuclear energy, utilities regulations, geothermal, wind, and overall administrative support." Pryor said D.O.E. had found no conflict of interest involved in hiring Booz Allen; but he added that the firm's own presentation to the Department of Transportation (D.O.T.) in 1979 "identified as 'some' of Booz's petroleum industry clients—Exxon, Gulf, Shell, ARCO, Texaco, Mobil, Standard Oil of Indiana, Standard Oil of California and over a dozen other oil companies."

But this was only half the story. Pryor continued:

In late 1978, Booz Allen also told D.O.T. of its work for OPEC oil companies, including the Government of Libya, Arabian American Oil Company, National Iranian Oil Company, Abu Dhabi National Oil Company, and Sonatrarch [Algerian National Oil Company].

As explained to the D.O.T. by Booz, this work was by no means limited to incidental or narrowly technical assistance, but included "defining long range objectives, plans and organization structure" (Sonatrarch); "long range planning" (Iran); "assistance in organization of hydrocarbon industries" (Libya); and a "broad range of technical and management consulting" (Abu Dhabi).

The Oil Stonewall   39

The investigation by Pryor's committee showed that other consulting firms filled similar triple roles as advisors to Big Oil, D.O.E. and OPEC. The effect was to put all three in bed together, with the consulting firms serving as the liaison for the cartel. The dominance of the multinational oil industry over the supposedly free institutions of government was indisputable.

Guttman probably best expressed the principles that were being violated when he told a Senate committee in 1979

The growth of the contract bureaucracy has been inconsistent with the principles of democracy, the letter and spirit of our laws, and tenets of good management. The contract bureaucracy has elevated invisibility, conflict of interest, and unlawful delegation of public authority to governing principles.

The Great Daisy Chain

5

Scandal

"I think we may have stumbled on the greatest criminal conspiracy in American history."

The words were those of Rep. Albert A. Gore, Jr., son of a formerly prominent U.S. senator from Tennessee, and there was Gore himself looking right at me as he spoke. It was just a brief flash on my television screen as I watched the NBC "Nightly News" on May 30, 1979. But the charge was sensational enough to jerk me upright.

"The greatest criminal conspiracy in American history" —what a story! I couldn't wait to see the headlines in the morning papers. But the next day I searched through *The New York Times*, the *Daily News*, *The Wall Street Journal*— and not only were there no headlines; I couldn't even find a mention of the story unless, indeed, it was buried in two or three paragraphs on page fifty-eight that I might have missed.

What was going on? What was this "greatest conspiracy" all about? A few telephone calls to Washington set me straight. Gore had not been talking about the long gasoline lines that had snarled up motorists from coast to coast in the spring "crisis," but about an older, sputtering-under-the-surface scandal known as the "Daisy Chain."

The Daisy Chain is best described as a Big Oil scam by which prices on controlled oil were written up through a series of phony paper sales and purchase agreements until the legitimate price was more than doubled. CBS in its "60

Minutes" telecast had nibbled at the fringes of the story in the fall of 1978, concentrating on one spectacular gouge in which oil supplied to Florida Power for its utility plants had been written up to cost millions of dollars more than it should have. CBS had lifted the lid on Pandora's Box, but there was a lot more inside the box. The Daisy Chain was not just an isolated ripoff of Florida Power: It was an industry-wide ripoff of the American people.

The basis for this bit of oil company chicanery was laid after the Arab oil embargo of 1973–74. In November 1973, Congress passed the Emergency Petroleum Allocation Act. The intended purpose was to protect the American public from price gouging and to ensure the equitable distribution of crude oil supplies. Price controls were imposed on old oil wells—in other words, those that had been drilled and were functioning before the Arab embargo. Newer wells drilled after the embargo were allowed to track the world price set by OPEC. The difference in the two prices represented literally billions of dollars. "Old oil" was fixed at a ceiling price of $5.25 a barrel; "new oil" was bringing $12 to $13 a barrel. Obviously, any unscrupulous manipulators who could spit at the law and transform "old oil" into "new" could reap fantastic illegal profits.

This was exactly what knaves high and low had been doing for years, but the Department of Energy, held in thrall from top to bottom by oil interests and their consultants, had blinked at the outrage. Then, in the summer of 1978, the scandal became too much for a young Texas lawyer to stomach. The maverick in the ranks of the D.O.E. enforcement division was a slender, thirty-three-year-old former Texas state prosecutor, Joseph D. McNeff. It was McNeff's testimony that had led to Representative Gore's outburst.

Since the media seemed as blind to the scandal as D.O.E., I telephoned McNeff on June 1, 1979. In a long interview, he explained the way the system had traditionally worked and the way the scam had developed.

Major oil companies were producing about 75 percent of the low-priced "old oil." This production had always been funneled directly from the wells in the field through pipelines to the majors' refineries. Oil drilled from smaller fields whose output did not justify expensive pipeline connections had been picked up by "resellers" who transported it to the refineries and were paid some 25 to 30 cents a barrel for their services.

Before the Arab boycott, McNeff said, there had been only 12 resellers active in the nation; now, in 1979 there were 592. What had happened, he explained, was that this new horde of resellers functioned as middlemen, taking over contracts that previously had linked producers directly to refineries. So the Daisy Chain was born.

The resellers job was to shuffle paper. They "sold" the "old oil" to each other. Each "purchaser" was entitled, naturally, to a legitimate profit; and so, by the time the "old oil" had passed through anywhere from six to ten hands, its price being written up at each step along the way, it emerged suddenly as "new oil" claiming the highest OPEC price. Instead of being worth just $5.25 a barrel, it suddenly became worth from $7 to $8 more. And when this kind of bonanza was being tacked on to hundreds of thousands of barrels a day, the loot was something the Mafia might have envied.

This system could not have worked, McNeff added, without the collaboration of the majors because the only way the resellers could break into the well-established chain of distribution was by using, or pretending to use, the majors' own pipelines. For this, obviously, they needed permission. The Daisy Chain had worked so smoothly, McNeff said, that some of the majors got greedy and set up their own Daisy Chains, laundering the production from their own fields through middlemen they had selected.

Since everyone involved was breaking the law, such transactions inevitably involved bribes and kickbacks. But these were hard to trace. Each reseller set up four or five dummy corporations, most of them chartered in the Bahamas—"And so, of course, you can't get their records," McNeff said. It was the oil industry's version of the Mafia's secret Swiss bank accounts.

When I talked to McNeff, he had no idea, I think, of the tight control exercised over D.O.E. by oil company links at the top and of oil company–connected consultants virtually running the department. What he knew was what he had seen in the field; there he had received a firsthand education about the mysterious way in which things that were supposed to happen never happened.

Auditors sent out by D.O.E. on the pretense of conducting investigations were kept on a tight leash before they even started. They were given specific instructions to look at just each isolated transaction. Handled this way, they could report that a certain amount of oil had been purchased by a reseller at a specific price, and then sold to a "purchaser" at a price just sufficient to justify a legitimate profit mark-up. Such figures indicated a clean deal. "We were saying, 'Hey, this is a chain,' " McNeff said. "But the auditors were never allowed to look at the chain aspects of it. There has been horrible, horrible mismanagement of the national office. When we would send up cases, we would be told, 'You've shown the chain, but you haven't shown any conspiracy because you haven't shown any payoffs.' Nobody wanted to find out." And without auditors free to ferret out the secrets hidden in the accounts of an entire chain of transactions, it was practically impossible to pin down the kind of payoffs D.O.E. pretended it needed to show a conspiracy.

In that conversation, McNeff told me something else that greatly intrigued me. He said that in the summer of 1978 *The New York Times* had sent one of its crack investigative teams to Texas. The reporters told him, he said, that they had been given carte blanche to run a thorough investigation, even if it took six months. Shortly afterwards, the *Times* was tied up by a New York newspaper strike, and all that ever saw the light of day was a watered-down, meaningless version that moved on the *Times*'s service wire. "The *Times* just dropped the ball," McNeff said. "I was very disappointed."

In the meantime, he had decided to blow the whistle in Washington. He contacted Representative Dingell's staff and told Michael Barrett, the committee counsel, and Peter Stockton, one of the ace investigators, everything he knew about the Daisy Chain. McNeff laughed a bit ruefully as he told me over the phone what happened next. "I made a mistake by telling a 'friend' of mine what I was going to do," he said, "and the word got back almost before I got to Washington. As soon as they found out, they took me off all the cases I had been handling. The only reason I wasn't fired, probably, is that Dingell wrote a strong letter to D.O.E. warning them not to dare to fire me, so I haven't been doing anything for the past year except reading testimony."

After talking to McNeff, I decided to go down to Washington. A second hearing on "the greatest conspiracy" was being held on June 4 by Representative Dingell's energy subcommittee and the House Subcommittee on Crime headed by Rep. John Conyers, Jr. Before this second hearing, I wanted to read through the transcript of the May 30 session. There I found in the record the verification of everything McNeff had told me—and more.

McNeff had testified that, when he was sent to Houston in 1977, he found "total chaos" in enforcement. Four reseller cases were then under investigation, and "it was almost impossible to find out what the oil companies were doing." He was given three of the cases to handle, but then the top auditor and attorney for the region were transferred to another divison. They were replaced, McNeff testified, by an acting head auditor, a man who had been suspected of taking a bribe from the Summit Gas Company, and by an acting head attorney who was a personal friend of Albert Alkek, "a key subject of our crude oil investigations."

These actions "paralyzed our efforts," he testified. And, as if to make certain that such efforts were so paralyzed that they could never be revived, the D.O.E. in December 1977 issued an "omnibus subpoena" for the records of all the major oil companies. "By this single act," McNeff testified, "the national office precluded any single attorney, investigator or auditor from demanding company records that they badly needed for their own . . . investigations from issuing a subpoena."

In further testimony, he described how he had come to Washington on June 5, 1978, to confer with Dingell's staff. On June 14, Barrett and Stockton flew to Houston. On that same day, McNeff testified, "the head of the F.B.I. office in Houston called me and berated me for going to Congress and asking what I was now going to be doing for a living. . . . He then sent agents to the D.O.E. offices and picked up the file cabinet [containing material on a case McNeff had been handling]."

This was only the first stage of the clampdown. On the morning of June 15, as McNeff and some of his associates were preparing to talk to Barrett and Stockton, "the general counsel of the Department of Energy, Lynn Coleman, and a deputy U.S. attorney general, identified as a Mr. Keeney, relayed messages to us that under no circumstances were we to let your [the committee's] investigators see any files or discuss any of the cases we were working on. If we did, we would be subject to prosecution."

This, clearly, was a brazen muzzling of D.O.E. enforcement agents to keep them from talking to congressional representatives—and to keep the whole scandal swept under the rug. The only conclusion to be drawn from McNeff's testimony was that D.O.E. was more interested in prosecuting anyone in its ranks who dared to talk than it was in prosecuting violators of the law.

There was another significant aspect to this threat. One of the officials who had made it, according to McNeff, was

D.O.E. general counsel Lynn Coleman. Coleman had been a law partner in the large and influential Houston firm of Vinson, Elkins, Searls, Connally & Smith. Another partner in the firm had been John B. Connally, former governor of Texas and presidential candidate, who had risen to riches through his services to Sid Richardson, one of the original Texas oil tycoons. Furthermore, according to Representative Dingell's staff, Coleman's former law firm had represented some dozen oil companies accused of price control violations. Only D.O.E., which repeatedly admitted to congressional probers that it never inquired into the associations of its officials and hired consultants, could have ignored the potential conflict of interest inherent in such a link between its general counsel and the oil interests of his former law firm.

McNeff's further testimony made it clear that drastic action had followed closely on the heels of the Coleman-Keeney threat of prosecution. He testified, "On June 19, when the D.O.E. inspector general flew from Washington to Houston to talk to me, all my files were sealed and locked up in another room. I was barred from working on any criminal cases, and attorney John Jensvold arrived from Washington to replace me."

Of special interest to me in assessing McNeff's credibility was the substantiating testimony given at this same May 30 hearing by two other D.O.E. enforcement officials from Texas. They were Herbert F. Buchanan, Jr., deputy district manager for enforcement in the southwestern district, and F. Edwin Hallman, Jr., regional counsel for Region 4 and chief enforcement attorney for the southwestern district.

Buchanan testified that one of his auditors had told him that Uni Oil, one of the suspected Daisy Chain operators, had built an entire new refinery *for cash.*

Hallman declared, "I have seen cases delayed for months and years. I have seen extreme problems that I, in fact, could not deal with, other than to register my objections."

Hallman had been especially shocked by one direct order from F.E.A., the oil-ridden predecessor of D.O.E.

There is a memo on file from the then head of the enforcement arm of F.E.A. [in] which he said *we will not hire criminal investigators to do criminal investigations.* And that was sort of like the telegram I got once saying there would be a fire drill at headquarters. I could not believe that it had been put on paper to the extent that it had. [Italics added.]

Hallman had also testified about a specific case involving a $150,000 payment to a former vice president of Gulf Oil. Gulf had access to a large amount of crude "of very fine quality," and another corporation, Nepco, wanted it. The vice president, who had the power to approve the transfer of oil, was paid $75,000 by Barko, a Bahamian corporation, 65 percent of whose stock was owned by Ncpco. After the oil was diverted, the Gulf vice president resigned, was paid another $75,000—and then signed a lucrative consulting contract with the firm to which he had transferred the oil. The case was kicked around for two and a half years by D.O.E. and was then sent to the Justice Department—but Justice said nothing could be done because the statute of limitations had expired, barring prosecution after the lapse of so many years.

Representative Gore had been incensed. "I just think it is fairly incredible . . . ," he said. "We have another case of Mobil interrupting its traditional patterns of delivery and inserting a dealer in the chain. The oil still moves along the same pipeline, but the paper chain now has a reseller in it, and the old oil becomes new oil."

Gore had then asked McNeff, "How many oil companies in your opinion are involved in these schemes?"

MCNEFF: I would say that almost all of them are, if not all of them.

So the stage was set for the confrontation of June 4, 1979. Dingell, Conyers and members of their committees faced an array of government witnesses: Deputy Attorney General John C. Keeney; Richard M. Fishkin, in charge of the fraud section in Justice's criminal division; and three F.B.I. agents—Dana E. Caro, chief inspector of the planning and inspection division; Francis M. Mullen, Jr., deputy assistant of the criminal investigation division; and Joseph E. Henehan, section chief of the white-collar crime section.

What quickly became apparent to me as I listened to the testimony was the sharp contrast between the conduct of these official witnesses and that of McNeff, Buchanan and Hallman. The D.O.E. enforcement officials who had worked in the trenches had given clear, positive, forthright testimony; here, at the top, it was as difficult to get a straight answer out of the official spokesmen as it had been for Energy Commissioner Jacobson in his confrontation with the oil industry's executive types.

It was obvious that Joseph McNeff was not the most popular man in official Washington, and government witnesses, when they got their chance, dismissed him with supercilious contempt and open hostility. They suggested McNeff was "paranoid." Keeney: "There is absolutely no evidence we have to support Mr. McNeff." Caro: "Mr. McNeff has gut reactions; he has suspicions, reactions, but no evidence." Rep. Michael L. Synar, a member of Dingell's committee, interrupted with a tart reminder, "We're not just talking about McNeff, but Hallman also." To this, there was no reply, just retreat into a kind of sullen silence.

The questioning became more intense. McNeff in his original testimony had mentioned Project X, the file that had been whisked out of his office just as Barrett and Stockton were landing in Houston. Project X, it developed, involved a major oil company, Conoco, and its relations with a number of resellers. This was the story:

Felonious activity involving some Conoco executives and resellers began in 1973. A double set of books was kept. Transactions eventually involved $1.4 million in price write-ups. High-up officials in Conoco became concerned and ordered an in-house investigation. After this, the company disclosed its findings to a representative of F.E.A., D.O.E.'s predecessor. Apparently appalled, the F.E.A. official appealed to the Justice Department for help.

Barrett, the counsel for Dingell's committee, put into the record this chronology: Conoco reported its guilt to F.E.A. on March 17, 1977; F.E.A. referred the case to Justice in July 1977; nothing happened until March 1978, when an attorney newly assigned to the case had been given "a wrong file"; it took another month for the attorney to get "the right file"; then he decided he had to redo the investigation. Finally, in July 1978, with the statute of limitations to expire in twenty-four hours, Justice made a plea-bargaining deal with Conoco.

Barrett, a sour look on his face, made it clear he found it decidedly odd that the Project X file, which had been lying around in the Houston D.O.E. office for nine months, was spirited away just as he and Stockton were arriving in Houston. The F.B.I. witnesses had an explanation: The file, you see, had been removed just at that time because the U.S. attorney had requested it for use before the grand jury; sorry, it just happened to have occurred that way. Barrett wasn't mollified. He pointed out that it was another three weeks before the U.S. attorney took the case before the grand jury. So what was all the rush on this particular day? And how had this all been allowed to happen?

"There was a slippage," Keeney acknowledged. The case had been assigned first to an attorney who was so burdened with other matters that he hadn't had time to handle it.

GORE: Why was it assigned to someone who didn't have time to handle it? . . . How long did it stay in his hands before you found out he didn't have time to handle it?

It had taken eight months to make this discovery, Keeney conceded; that accounted for the gap between July 1977 and

March 1978. The wrong-file–right-file foul-up followed, causing further delay.

GORE: You finally got a conviction on the very day the statute ran out. Billions of dollars are involved in these cases, and the American people want vigorous prosecution.

Dingell emphasized that, in this instance, Conoco had turned itself in—and still, despite Conoco's cooperation, the case had been almost allowed to die.

GORE: If under these circumstances, it takes five years to get a conviction, it's no wonder so many others get off scot-free— particularly when others are not as cooperative as Conoco was.

Gore had another bone in his throat. It involved Hallman's May 30 testimony about the $150,000 payment to a former vice president of Gulf Oil in a reseller scheme. Why had nothing been done in that case?

KEENEY: The case is closed as far as criminal prosecutions are concerned. It is open as far as civil action is concerned.

Fishkin, his Justice Department colleague, interjected that Justice had determined "the case had no prosecutorial merit." Correcting Hallman on one point, he made it clear that the case hadn't been dropped because the statute of limitations had expired. It had been referred, Fishkin said, to the U.S. attorney in New York; a grand jury had heard the case and refused to indict. Considering the well-established chain of evidence that would indicate a $150,000 payoff, one has to wonder just how vigorously the case was presented to the grand jury. Fishkin contended, "We must be able to prove guilt beyond a reasonable doubt. If we can't, the case has no prosecutorial merit." It began to seem as if the Justice Department had excuses for not prosecuting anyone for anything.

The impression was reinforced when the joint congressional committee hit the official witnesses with an account of a pyramiding scheme that its own investigators—not the F.B.I., not Justice—had uncovered. It involved a firm called La Gloria Oil and Gas Company, which made its headquarters out of P.O. Box 2521, Houston, Tex. A document obtained by congressional investigators showed that on November 7, 1974, letters from La Gloria were sent to Frank English, Pedco Oil Company, Houston; Bob Johnson, Conoco Oil Company, Houston; Lee Moore, Energy Marketing Company, Abilene, Tex.; and Bob Whitson, Gustafson Oil Company, Wichita, Kan. The letters all read:

Gentlemen:

The following *book transfer* [Italics added.] has been agreed to among all companies concerned and will be effective in October business.

Product: No. 2 fuel oil.

Volume: 25,000 barrels.

Transfer sequence: La Gloria to Energy Marketing to Pedco to Gustafson to Conoco to La Gloria, Tyler, Tex.

The 25,000 barrels of oil, in other words, never left Tyler. The scam began with La Gloria and ended with La Gloria; but the intervening paper transfers had boosted the price of the oil at every stage in the chain. As far as volume went, this was not an especially large transaction, but the sequel showed it had been a highly remunerative one. A subsequent letter from La Gloria notified one of the conspirators that a check was being enclosed for $32,840 as his share of the "profits."

When Deputy Attorney General Keeney and the F.B.I. sleuths were confronted with this documentary evidence, they protested that this was the first they had ever heard of the La Gloria case. Dingell pointed out that the case was nearly five years old, that the statute of limitations was about to expire. Where had Justice been all this time? Keeney argued that the Justice Department would have had no way of knowing about the case unless it was referred to it by D.O.E. Dingell snapped back that the committee had been informed the case had been closed over the protests of D.O.E. auditors.

Keeney attempted to show that the Justice Department was now prosecuting more vigorously. Originally, he conceded, there had been a breakdown in communications with D.O.E., but now that had all been straightened out and enforcement efforts were going to be much better in the future.

Representative Synar was getting restless. He took up questioning. It went this way:

SYNAR: What are we talking about? Are we talking about amounts in the billions of dollars?

KEENEY: Yes, we are.

SYNAR: To the best of your knowledge, are we talking about a limited ripoff or are we talking billions—and in all fifty states?

KEENEY: I'm not a good person to answer that question.

SYNAR: Is it possible for this to exist without the collaboration of the major oil companies?

KEENEY: I don't think I can answer that.

SYNAR (plainly incredulous): You can't answer *that*?

Chairman Conyers was becoming disenchanted with such nonanswers. When Keeney protested that he did not have any way of estimating the magnitude of the fraud, Conyers shot back, "There isn't any other agency of government that is sitting on top of the largest criminal conspiracy in American history, is there? . . . We are talking about an industry that is costing each and every American to pay more."

Conyers tried a new tack. Was the name John C. Sawhill familiar? he asked Keeney. The deputy attorney general looked puzzled, as if the name was entirely foreign to him, but, as Conyers pressed, he finally conceded that he had a vague recollection of having heard the name somewhere. Keeney's vagueness was interesting because Dr. John C.

Sawhill had been chief of enforcement for F.E.A., and Joe McNeff in our first telephone conversation had told me that Sawhill had understood the whole old oil–new oil ripoff as early as December, 1974.

Conyers now produced a copy of the *Washington Post* for December 17, 1974. He read from a front-page article describing a press conference Sawhill had held. Sawhill had announced that he was forming a special staff of thirty investigators for what he called Project Escalator. This task force, Sawhill announced, was going to investigate five categories of oil industry crimes, including bribes, kickbacks and phony paper transactions to write up the cost of oil. Did Keeney recall any of this? Conyers wondered.

Keeney didn't. Well, Conyers pointed out, Sawhill was not just some average citizen walking in off the street with a tall story; he was an important federal official, the chief law enforcement officer of F.E.A. Wouldn't Justice have been aroused by such sweeping charges? Wouldn't it have jumped into action?

It would have, Keeney said, if Sawhill had sent it anything "specific" instead of just holding a press conference to get publicity. Well, Conyers pressed, *had* Sawhill given Justice any specifics? Keeney didn't know, but Justice wouldn't have paid any attention to him unless he had. "Not even though he was the chief federal enforcement officer in the field?" Conyers wondered in amazement.

Not mentioned at the hearing was one significant fact: the ax fell on Sawhill in record-breaking time. Just two weeks after his December 17, 1974, press conference, he was forced to resign. Project Escalator died with him. Evidence that he had wanted to pursue was discarded in dust-gathering files like those that Joe McNeff had found on one occasion in Houston. And the racket continued for years, unhindered by D.O.E., the F.B.I. or the Justice Department. Those who paid were the American people—rooked out of billions of dollars by the Daisy Chain write-ups—and the few men of conscience who had tried to do something about it, Sawhill, Brock Adams, Joe McNeff.

It had been, as such things go in Washington, a fairly explosive hearing, and the hearing room had been packed with reporters and researchers, all furiously taking notes. It was a mystery to me then—and I was to have the same experience many times later—exactly what become of the fruits of this furious activity. For the press in New York, and the media generally as far as I could discover, remained blind to the issues that had been raised.*

I had been left, however, with the feeling that the two hearings of May 30 and June 4 had merely lifted the lid, that much more remained to be uncovered and told. And in this I was right. Dingell's staff of researchers kept digging.

A subsequent staff memorandum, a copy of which I have, shows that Dingell's investigators, unlike the F.B.I. and Justice, diligently interviewed executives on all sides of the Daisy Chain issue. It shows that they had to wade through packs of lies to get at kernels of truth.

Paul Bloom, special counsel in D.O.E.'s Special Investigation group, told Dingell's office that he had referred as

---

* After the June 4 hearing, there were some glaring examples of how *not* to write the real story. *The New York Times* devoted most of a column on an inside page of its business section to an exculpation of the Justice Department. The story for the *Times* was Keeney's statement about the number of investigators Justice was *now* training and what it promised to do. In the sixth paragraph, there was a passing reference to Joe McNeff; there was no reference anywhere in the story to the threats to McNeff, the testimony of Hallman, the tough questioning of Keeney, the Sawhill episode with all that it implied. The Associated Press wire story out of Washington didn't make even a mention of McNeff and devoted itself entirely to official explanations of how errors of the past were going to be corrected. In neither account was there any indication of the possible involvement of major oil companies in the Daisy Chain, whose very existence was so sketchily described that it appeared to be a minor matter of little significance.

many as sixteen or seventeen cases involving criminal activity on the part of the majors to the Justice Department for prosecution. Justice officials insisted they had received only two or three such referrals, and they admitted that they had initiated no investigations of their own.

"By the time of the hearings," the memorandum says, "D.O.J. informed the staff there had been no convictions, indictments or no grand juries had been impaneled to investigate potential criminal activity on the part of the majors."

After the June hearing, Dingell's investigators went to Houston to follow up on Daisy Chain leads. They interviewed officials of Gulf, Exxon and Shell.

Spokesmen for all the majors claimed at first that they had refused to deal with any resellers who came into the picture after 1973. They belittled the role of the resellers, but Dingell's investigators noted that "this makes little sense in that the majors control as much as 75 percent of the old oil and the resellers needed this old oil for their flip schemes."

The staff memorandum continues.

They [the majors] admitted that they had knowledge that the resellers' schemes were operating; however, they claimed that they had no proof so they didn't report their information to D.O.E. or D.O.J. . . .

They advised the staff that most of the resellers were created by former officials of major oil companies—estimates as high as 60 percent. . . .

Gulf said that resellers were set up by independent producers to circumvent price control by providing kickbacks in the form of money or services.

Gulf [Jim Streit, former manager of crude oil] talked to Carl Corallo of D.O.E. in 1974 about the reseller problem. He never heard from Corallo again.

Through interviews with D.O.E. enforcement officials in Houston, Dingell's staff "obtained copies of pipeline transfer letters indicating ownership of crude oil in the major pipelines." The majors had dragged their feet for nine months before delivering this information. What it showed was summarized in the staff memorandum in these words:

This crucial data revealed not only the illegal flipping of the old oil by resellers but also a substantial involvement in the transactions by a number of major oil companies.

Contrary to the statement by the majors that they refused to deal with the resellers, *the pipeline data revealed active participation in the schemes by the majors.* [Italics added.]

Yet nothing had been done, nothing was ever done, to prosecute any of the major oil companies or their officials for their involvement in the Daisy Chain. One of Dingell's investigators spoke with disgust of the performance of the Justice Department. "They are a bunch of eunuchs over there," he said.

What had all of this officially permitted crookedness cost the American consumer? No one will ever be able to total up the full cost, but a G.A.O. report that had not been publicly released at that time indicates the magnitude of the ripoff.

F.E.A. statistics, collected at the wellhead and then at the refinery, indicated that over 300,000 barrels per day of old crude were converted to new crude during the month of March, 1977. As of September, 1977, the rate declined to over 139,000 barrels per day. At the lower September rate, D.O.E. stated that over $800,000 per day was being unlawfully taken from consumers. At the $8 per barrel difference, the March, 1977 conversion rate indicates *over $2.6 million per day may have been taken.* . . . [Italics added.]

Even if the Daisy Chain operated for only twenty days in this month of March (assuming that even thieves take weekends off), the cost to the American consumers would have to come to more than $52 million. In just one month! And the Daisy Chain had operated month after month, year after year, with the covert blessing of officials who were supposed to represent and protect the American people.

Thus "the greatest criminal conspiracy in American history" died without a whimper—without the arrest and conviction of any of the major conspirators involved.

6

The 1979 Gasoline

"Crisis"

Pandemonium swept the gas stations in New York City on the weekend of June 8–9, 1979. Frenzied motorists queued up in such long lines that their cars sometimes blocked traffic and forced the closing of stations that were pumping gas. In Brooklyn, Fritz Boutain, twenty-nine, was waiting in just such a frustrating line when another driver tried to wedge in. There was a fender-bender, followed by a fist fight. Boutain drove off the line-crasher, but a few minutes later the man returned, reached through the open window by the driver's seat and stabbed Boutain in the heart with a hunting knife. As two women passengers in Boutain's car screamed in horror, he tumbled from the car and died; his killer fled.

The press noted that it was the second gas-line killing in New York in ten days.

This irrational violence was the result of a gasoline "shortage" that had swept the nation with the suddenness of a Kansas cyclone. It had started first in California. There the pumps had begun to go dry in early April. In California, of all places! In California, where the huge new production from the Prudhoe Bay oil fields in Alaska had been so inundating that there had been suggestions some of the oil might have to be shipped to Japan. Yet it was here in California that motorists suddenly found they couldn't get gas for their cars. Stations stayed open for only a few hours and limited sales to regular customers. The familiar pattern of miles-long waiting lines and of frustrated drivers jockeying for position began to develop.

By May 21, the crisis in California had eased as suddenly as it had arisen. Gasoline became miraculously available again. At some stations, there were open pumps—and no lines of cars waiting for a fill-up.

But elsewhere in the nation, there was chaos. Independent truckers went on strike in June. They blocked major highways in protest against the scarcity of diesel fuel, its soaring price and the fifty-five-mile-an-hour speed limit. Deliveries of perishable produce were delayed. Much of it was spoiled, and some farmers simply dumped their rotting harvest out on the highways and plowed under the rest of their fields.

A nationwide truckers' strike on June 22 induced President Carter to order the F.B.I. to provide assistance in keeping the highways open. Shootings and other incidents of violence were reported in Rhode Island, Connecticut, Pennsylvania, South Carolina, Kentucky, Oregon, Louisiana, Mississippi, Utah and Oklahoma.

Motorists in the East were as enraged and some of them, at least, as prone to violence as the truckers. The plague that had begun in California had swept east to Washington and New York. Even as California began returning to normal, earlier creeping shortages in the East became a major gasoline drought. The Memorial Day weekend and the first ten days of June were crisis times, marked by panic, rising anger and violence. In the Bronx, a free-for-all broke out among drivers of gypsy cabs at one gas station. Knives, crowbars, two-by-fours, and jack handles were wielded in what police called "a real donnybrook." In the midst of the melee, a .38-caliber revolver came into play. Two persons were wounded.

And gasoline prices soared. Motorists, unable to get gas when they needed it, were willing to pay almost any price for it when they could get it.

Regular leaded gasoline that had been selling for 59.9 cents a gallon, unleaded for some 6 or 7 cents more, when I went into Newark for Jacobson's hearings in December, 1978, only five months previously, now cost almost a dollar a gallon. On the Memorial Day weekend, one station owner in Queens boosted his unleaded price from 93 cents to $1.30 a gallon.

And this was only the beginning. No longer was there "a gasoline price war on," no longer did this liquid gold have to be sold for less than the permitted ceiling price; now the big leap could be made through the ceiling as those "banked credits" from the gasoline-war past could be tacked on to the top ceiling price.

The *Daily News* recalled that Energy Secretary James R. Schlesinger had assured the public in February that prices would rise no more than 9 *cents a gallon in the next two years*. It was an official estimate that now looked as ridiculous as F.E.A.'s assurance that heating oil prices, at the very worst, would rise no more than 1 cent a gallon after decontrol. The fraud in such official assurance so swiftly belied by events was obvious.

I flashed back to Shell's abortive attempt, aided and abetted by the other majors, to create a shortage-panic situation the previous December. I recalled my favorite gas station owner's prediction *then* that "you're going to have lines waiting at the pumps again." Vivid was Joel Jacobson's January assessment that, despite talk of shortages, "there is no real shortage. It is a contrived situation to drive prices up."

Now it had all come to pass—in the most mysterious of ways, as if the whole mess had been carefully orchestrated by a public relations Svengali.

The most inexplicable feature of the whole "crisis," the one event above all others that made it smell like rotten eggs, was the manner in which the gasoline drought had hit *first* in California and along the West Coast. As late as February

16, 1979, *The Wall Street Journal* had run a lead story that began:

> Stirred by the cutoff of oil from Iran, a nervousness reminiscent of the days of the Arab oil embargo has begun to spread over the country. . . . Yet, in one part of the country, the West Coast, there is too much oil. West Coast refineries are able to use only about two-thirds of the 1.2 million barrels a day of Alaskan crude oil flowing from the Alaska pipeline into tankers at Valdez.
>
> For the time being, the excess is being shipped through the Panama Canal to Gulf Coast refineries.

Yet, like a curtain being drawn at end of a play's Act I, the excess oil vanished—and California pumps began to go dry. The Washington state attorney general's office subsequently investigated the mystery. The report that it issued was the most caustic to be turned in by an official investigative agency. This is a summary of its chief conclusions:

- First, the report found that "competition was absent from the events surrounding the 1979 shortage." It added that "the oil companies at the very least have been the enthusiastic beneficiaries of OPEC policies and D.O.E. mistakes."
- While D.O.E. technically was responsible for gasoline allocations to various parts of the nation, "the requests to commence allocations, the statistical data on which the allocations have been based and calculations of the details of the actual allocations all have originated with the companies themselves."
- These allocations, cutting gas delivered to service stations to between 70 and 90 percent of 1978 volume, had "the severest impact in and around large cities." (Where, of course, shortages could generate the greatest publicity and create the worst panic.)
- Consumption in the first five months in the five Western Pacific states (known as Pad V) was 1.8 percent higher than in 1978, *but gasoline output was 4.8 percent higher.*

• Finally, "it is difficult to explain the observed shortages
on the West Coast, remembering what occurred in 1973,
without suspecting collusion among the major oil companies. In a competitive petroleum market with the
available crude and excess refinery capacity indicated by
Pad V statistics, a cutback of allocation by one company
should have resulted in increased sales, market share
and profits by other companies."

There had been no such competition; there had been only
escalating prices and usurious oil company profits. In April
1979, just as the gasoline "crisis" was engulfing the nation,
the major oil companies released reports on their first-
quarter earnings. Exxon's profits had soared 37 percent over
the incredible first quarter of 1978; Indiana Standard's, 28
percent; Occidental Petroleum's, 174 percent; California
Standard's, 43 percent; Ohio Standard's, 303 percent;
Conoco's, 343 percent; Mobil's, 81 percent; Texaco's, 81 percent; Gulf's, 61 percent; Shell's, 16 percent; Phillips's, 4
percent; Amerada's, 279 percent; Marathon's, 108 percent;
and Cities Service's and Getty's, both 42 percent.

Such enormous profit increases (Exxon's $955-million
profit in just one quarter shocked some observers, who
couldn't know, of course, that Exxon would soon be posting
a one-quarter profit of *$1.9 billion*) incensed the American
public. The fact that Big Oil was wallowing in such unheard-
of profits at the very time the average American was coping
with long gas lines and higher prices made the situation
seem almost obscene. Polls showed that two-thirds of the
American people didn't believe the gasoline "crisis" was real;
they felt it was just a Big Oil ploy to increase the size of its
gouge. Even Sen. Howard H. Baker, Jr., in a nationally televised appearance on May 13, remarked angrily that the big
oil companies were asking for divestiture unless they re-
formed. It was an uncharacteristic outburst from the con-

servative Baker, now the Senate majority leader, and it was the last such outcry the nation was to hear from him.

Even so, Baker's momentary tweak of conscience and concern was refreshing compared to the performance of Jimmy Carter in the White House. Carter bought without an apparent qualm the oil companies' two-fold explanation of the long 1979 gas lines. Motorists hadn't been conserving but had been using more gas, especially in the mild fall of 1978, the oil companies said. On top of this, the revolution in Iran that had deposed the Shah had cut off Iranian oil shipments, creating a shortage.

The first argument was shot down by the industry's own lobbying arm, the American Petroleum Institute (A.P.I.). It reported on May 16, 1979, that the nation's total demand for gasoline and other refined oil products had been the lowest in April of any month in nearly two years. So how did we suddenly have a shortage? Well, A.P.I. said, domestic production had fallen off as the result of bad weather, and the nation's refineries were running at only 85 percent of capacity instead of a normal full-production rate of 91 or 92 percent. The bad weather argument would later be discounted by official sources, and refinery runs were something that, obviously, were under oil company control. The result, however, was that oil companies delivered less gasoline to their dealers in March and April.

Even this industry mouthpiece damaged the alibi of the Iranian shortfall. It conceded that the nation had imported 4.3 percent more crude oil and refined products in the first four months of 1979 than it had in the same period in 1978. Though the Iranian shortfall was supposed to have cut imports by 600,000 barrels a day, A.P.I.'s figures showed that, in the month of April alone, imports were 7 percent above the figure for the previous year.

A.P.I.'s own assessment, when one analyzed it, indicated that the public's suspicions about a contrived shortage were justified. There was great skepticism about bad weather having caused a falloff in drilling, but even more obvious was A.P.I.'s own admission that there were plentiful crude oil supplies. The big gap, from its own statement, seemed to have been caused by a cutback in refinery output (sometimes refineries were running as low as 84.1 percent of capacity) and the subsequent reduction of supplies to the nation's gas stations—both factors controlled by the oil companies.

This industry-supplied data conflicted with the Carter–Big Oil script of a disastrous Iranian shortfall. Carter kept crying at every opportunity that the shortage was real. But Jack Anderson, the Washington columnist, published excerpts from secret White House meetings indicating that President Carter himself had deliberately cut back gasoline supplies to keep his pledge to other industrial nations that the United States would reduce oil consumption by 5 percent. At a May 7 meeting, just as motorists were tangled for miles in long gas lines in California, Carter told his Cabinet, "Our priority will continue to be some heating, agriculture and emergency needs over highway driving. . . . There will be less gasoline, and it will cost more."

This high-cost scenario was central to the Carter administration's efforts to cope with the energy problem. Jerry Ferrara, the outspoken director of the New Jersey Retail Gasoline Retailers Association, described in a television appearance on July 2 how he and his associates had "pounded on every door" in Washington seeking the adoption of a more sensible policy. He said he had met face to face with Secretary Schlesinger, and he added, "[Schlesinger] said that if gasoline got up to two dollars a gallon by 1981, the American people would have to conserve. And then he walked out of the room."

The reek of a Big Oil–Administration rat was becoming so pervasive that both Senate and House committees tried to find out what the devil was happening. They were no more successful in getting straightforward answers than Joel Jacobson had been. After one session with major oil executives, Senator John Durkin exclaimed in exasperation, "Getting the truth is like trying to nail Jell-o to the wall."

In the House, Rep. Benjamin S. Rosenthal's Subcommittee on Commerce, Consumer and Monetary Affairs fought several tough rounds with D.O.E. officials. The preliminary round took place on June 11. Alfred Dougherty, director of the Federal Trade Commission's Bureau of Competition, testified that there had been no falling-off in oil imports as a result of the situation in Iran. He added, "It appears that some crude oil is being stockpiled." He said his conclusions were based on publicly available information; he had tried to get specific data from D.O.E.—but D.O.E. had refused to cooperate.

This was followed by the appearance of John F. O'Leary, deputy secretary of energy, who peddled the Carter line that OPEC was responsible for all our troubles. "We are absolutely at the mercy of the international oil cartel," he said. He admitted that Energy Department figures showed "the amount of gasoline produced from available crude oil during the first five months of 1979 was very similar to the percentages for 1978 and 1977." What about Dougherty's assertion that D.O.E. wouldn't release its figures? O'Leary insisted that Dougherty had never asked for the information. "We have searched the records; we have no request for data," he said.

The third round in this encounter pitted Rosenthal in a knockdown bout with O'Leary. Though O'Leary denied the D.O.E. had tried to block a Federal Trade Commission probe of the gasoline shortage, he refused to release financial records of the oil companies to prove his assertion that there had been no conspiracy. D.O.E., he said, had "an agreement" with the oil companies not to dislcose any data it got from them. This was too much for Rosenthal.

"You have to cooperate!" he roared angrily. "You are an

employee of the United States government—you are not a czar appointed by another czar! You have lost the credibility of the American people and the Congress. You are on the verge of bringing down a presidency with this kind of attitude."

The sequel was described by Rosenthal in the *Congressional Record* of June 29.

Following the hearing of June 14, Secretary of Energy Schlesinger admitted that we had more than adequate inventory stock and that he would undertake to use the Department of Energy's allocation authority to urge recalcitrant refiners in the direction of serving the consuming public. . . . On Thursday, June 24, Secretary Schlesinger reversed his stand, expressing fear that the U.S. multinational companies might retaliate by withholding oil from the United States. Thus, it becomes obvious that much of the blame for the current gasoline shortages must also be ascribed to deliberate actions by the oil companies and the Department of Energy. This "blackmail" threat by the U.S. multinational companies that ship crude oil to Europe instead of to the United States calls for a vigorous response by this nation.

This abject surrender to "blackmail," as Representative Rosenthal so aptly put it, was clear proof, if any more were needed, of just who was the "czar" in America. Big Oil was on the throne, and the government of the United States was literally its servitor.

The result was the beginning of what Rosenthal had called "bringing down a presidency." In these same first six months of 1979, while Big Oil was playing games with the nation's welfare, the economy went into the beginnings of the tailspin from which it had not yet recovered. BURST OF INFLATION HIGHEST IN 4½ YEARS, PERILING GUIDELINES, read *The New York Times* headline of March 24, 1979. February's prices had leaped 1.2 percent, giving the nation an annual rate of 15.4. FOOD AND FUEL COSTS LED INDEX IN NEW YORK AREA, the *Times* explained in another headline.

A 60-percent increase in the price of jet fuel within a

year had sent the airline industry into the red. Huge losses were reported: United and American lost more than $60 million each in the first quarter; Pan American had a record first-quarter loss of $74.9 million.

"Uncertainty surrounding the availability of gasoline," a Detroit spokesman said, had sent the automobile industry into the skid that still continues. MID-MAY AUTO SALES OFF SHARPLY, the *Times* noted on its business page. The Big Three auto makers reported sales down 25.7 percent in the first ten days of May. They were to continue to plummet, with Chrysler saved from bankruptcy only by a federal bail-out, with Ford losing $164 million in the first quarter of 1980 and being forced to skip its first-quarter dividend in 1981.

The dreaded word "recession" began to creep into the headlines. By May 1, 1980, the *Washington Post* was head-lining, U.S. REPORTS FRESH SIGNS OF RECESSION. The un-employment rate was heading for 8 percent.

Such were the fruits of policy that blatantly benefitted only Big Oil at the expense of the nation and the welfare of its people. In time it would, indeed, bring down Jimmy Carter's presidency; but, in 1979, though the impending disaster should have been obvious to any thinking man, Jimmy Carter clung to his Big Oil script with the fanaticism of a drowning man holding onto a life raft.

He called the charges that 1979's shortages had been contrived by Big Oil to drive up prices false. The American people were just looking for a scapegoat, he said, instead of facing up to energy problems as they should. He was calling for the decontrol of oil prices (something Congress wouldn't permit) and a "windfall profits tax" that would drain off any excessive profits made by the oil industry. He remained blind to the obvious fact that these profits already were excessive.

"We are still feeling the effects of March [the Iranian shortfall]," he said in another statement. "It takes two months to transport oil from Iran to the United States and additional time to get that crude oil through the refining and distributing system. This has contributed to making late April and May the low points in gasoline availability."

Finally, in early July, the president retired to his mountain retreat at Camp David in the Maryland hills for ten days of soul-searching about the energy crisis and what to do about it. After those ten days on the Mount, he came down to Washington on July 15 and delivered a nationally televised address that was three-quarters corn pone and one-quarter design for action.

In the long corn-pone preamble, the president seemed to be blaming us, the American people, far more than he blamed himself and his administration. He was distressed, he said, by what he detected as "a crisis of confidence" in the nation that posed "a fundamental threat to American democracy." This national malaise of the spirit had kept us from uniting and working together for a better future. Too many of us now tended "to worship self-indulgence and consumption." We would have to change.

The villain at the root of all our problems was OPEC—just OPEC, not OPEC and Big Oil in partnership.

Our excessive dependence on OPEC has already taken a tremendous toll of our economy and our people. This is the direct cause of the long lines that have made millions of you spend aggravating hours waiting for gasoline. It's a cause of the increased inflation and unemployment we now face. . . .

The energy crisis is real. It is a clear and present danger to our nation. These are facts and we simply must face them.

When he came to the "facing" part of his speech, he clenched his fist and hammered it up and down to emphasize his points.

Point 1: I am tonight setting a clear goal for the energy policy of the United States. Beginning this moment, this nation will never use more foreign oil than we did in 1977. Never.

The glare in his eyes, the tightening of the jaw, the thump of the fist may have made this pledge sound like strong talk to the unitiated, but to those who knew anything about the crude oil problem, nothing could have been more ridiculous. Carter wasn't telling the American people that, in 1977, we imported a record-breaking 8.7 million of barrels of oil a day. We had never approached that figure since—and so his pledge not to do it again was meaningless.

The rest of his speech was devoted to proposals to make the nation more energy self-sufficient. He urged conservation as he always had. He called for electric utilities to cut their oil consumption by 50 percent through more use of coal. He announced "the most massive peacetime commitment of funds and resources in our nation's history" to develop synthetic fuels from coal and grain and the sun. He envisioned "20 percent of our energy coming from solar power by the year 2000." To finance this program, he called on Congress to pass a windfall profits tax to siphon off excess profits from the oil companies.

As Hedrick Smith wrote in an analytical article in *The New York Times*, the speech marked "an important turnaround from two years ago when, in setting out his first comprehensive energy package, the president concentrated almost exclusively on trying to force energy conservation through a variety of taxes and tax incentives." Now he was placing emphasis for the first time on greater energy production, much of it from renewable sources.

It was in the aftermath of this speech that I wrote for *The Nation* an article denouncing the entire 1979 gasoline crisis as phony. "How Big Oil Turned Off the Gas" was the title. The *Washington Post* bought syndication rights from *The Nation* and featured the article. And *that* caused one mighty ruckus in Washington.

The Phony Crisis

7

The great "oil crisis" of the summer of 1979 may well go down in history as one of the greatest frauds ever perpetrated on a helpless people. The truth is that there was no shortage of oil. This is verified by every responsible source. Indeed, solid statistics show that there was more oil available than there had been in 1978 when there were no gas lines, no murders of frustrated motorists—in a word, no "crisis."

That was the opening paragraph of an article that I wrote for *The Nation* in its issue of July 28–August 4, 1979. It was true then; it remains true today, despite the extreme measures to which the Carter administration resorted in an effort to prove it wasn't so. Those efforts included the ruthless muzzling of every dissenting, experienced voice within the administration; in the end, some heads would roll.

There had been ample grounds for suspicion before I wrote my exposé for *The Nation,* but the suspicion had never been supported with the kind of solid facts that I had. Here I should explain that I was helped greatly by a young intern working that summer in one of the senatorial offices. Peter Deutsch had been intrigued by the oil articles I had been writing, and he supplied me with hard facts he had gathered from official agencies. Without his help, I couldn't have written as strongly as I did.

Some background is essential to an understanding of what had happened. The record 8.7-million-barrel-a-day imports of 1977 had stemmed from fears by the major oil companies that the OPEC countries were going to boost prices drastically. OPEC, however, had not obliged in 1977— and so the oil companies were left with huge inventories. As a result, they began to draw down these surplus stocks in 1978, coming into 1979 with less oil in the stockpile.

The so-called Iranian shortfall in early 1979 thus came along at just the right time to lend color to the rumors of impending shortages that had been purveyed at the end of 1978. Only, as all reliable figures showed, there had been no real "shortfall." Iran supplied only about 4 percent of our imported oil. Increased production from Alaska, Saudi Arabia, Nigeria and other sources more than compensated for any loss from Iran. Indeed, as one quickly squelched U.S. Treasury analysis would later show, the oil companies in 1979 were actually *better off,* than they had been in 1978, both from the standpoint of supplies and the better grades of oil that made up for any loss from Iran.

Despite the drawdown in stocks in 1978, Department of Energy figures, based on those of the American Petroleum Institute—the only source D.O.E. had—showed that there had never been any danger of a shortage. Disregarding oil stored in the Strategic Petroleum Reserve's salt domes in Louisiana, the crude oil supplies available for domestic use showed this situation:

- At the end of 1976: 285 million barrels in reserve
- At year-end 1977 (the glut year): 339.859 million barrels in reserve
- At year-end 1978 (the drawdown year): 314.462 million barrels in reserve.

In other words, though the nation came into 1979 with a decrease of 7.5 percent from the start of 1978, it still had 10.2 percent more crude oil in reserves than it had at the beginning of 1977, the year of huge imports and stockpiling.

And it entered 1979 in far better shape than it had been in during 1976, when there had been no thought of a critical shortage.

How could this be explained? The petroleum industry and the Carter administration kept up a propaganda barrage about the Iranian shortfall as the cause of all our troubles. Lesser excuses included the contentions that bad weather had reduced domestic drilling in the winter of 1979 and that Americans had been driving too much, consuming more gasoline. Let's look first at the cornerstone of these defenses, the myth of the Iranian shortfall.

A Federal Trade Commission internal study, limited to gasoline alone, showed this comparison:

Barrels of Gasoline in Reserve

	(IN MILLIONS OF BARRELS)	
	1978	*1979*
JANUARY	195,920	197,271
FEBRUARY	199,048	207,054
MARCH	241,873	256,223
APRIL	202,983	249,384

Remembering that 1978 supplies were more than ample, representing the spill-over from the 1977 glut, it is startling to see that, in every month of the first quarter of the "crisis" year, there were more supplies on hand than had been available in 1978, when there was so much gasoline price wars were actually being waged. And, in April, when the gas drought began to hit California, the gasoline supplies were actually 22.9 percent *higher* than they had been in 1978.

In a memorandum dated May 30, 1979, Marc Schildkraut, a F.T.C. attorney, summed up the situation in these words:

The data indicates, among other things, that gasoline supplies were up by 4–8 percent, depending on the time period, over

1978's. Net supply of gasoline in April was particularly plentiful compared to the previous April (up by 22.9 percent). Significantly, however, every time period—month, quarter and third— shows increased supplies and no indication of a shortage.

Similar conclusions were reached in every independent study. The Central Intelligence Agency's *International Energy Statistics Review*, U.S. Customs Service reports on imports, International Energy Agency (I.E.A.) statements —these and other studies all agreed that free world crude oil production in 1979 had been higher than in 1978; that imports to the United States had been higher; that, in other words, there was no legitimate reason for the gasoline-crisis panic that had swept the nation.

C.I.A. figures showed that free world production, expressed in thousands of barrels daily, rose to 46,515 in the first quarter of 1979 compared with 46,305 in 1978. The C.I.A. study showed that imports in the first five months of 1979 outstripped those of the same period in 1978. During the first three months, imports averaged well over 8 million barrels a day, and in April and May they were only slightly below that figure. By contrast, in 1978, imports reached the 8-million-barrel-a-day figure in only two of the first five months, and in the other three months they trailed considerably behind the 1979 import figures.

U.S. Customs records told a similar tale. These should be highly reliable because Customs conducts its own check on incoming cargoes. Yet Customs figures, as filed with the Census Bureau, showed that imports for the first five months of 1979 had increased 10 percent over those for the same months in 1978.

International experts with whom American oil companies filed reports could find no cause for a crisis. In late February, Dr. Ulf Lantzke, director of the I.E.A., replied to a Royal Dutch/Shell spokesman's assertion that the shortfall in world supplies was as bad as that caused by the Arab oil embargo of 1973–74. Dr. Lantzke said, "We do not think that there is any cause for panic," and he added that oil supplies were assured for at least the first quarter of the year.

In the House of Representatives, Representative Dingell's energy subcommittee staff had been making its own investigation. It came up with some devastating conclusions. In a memo dated March 21, 1979, entitled "Schlesinger's Gulf of Tonkin," the staff reported:

A staff review of numerous D.O.E. documents revealed that, according to data compiled by the International Energy Agency, U.S. imports during the month of February surged by one million barrels per day, despite a significant decrease in Iranian imports. At the same time, D.O.E. reports to this subcommittee and the public were showing no increase at all. Internal D.O.E. memoranda reveal that officials were aware of the discrepancy and attempted to reconcile it by phoning the oil companies. When the discrepancy could not be resolved, no attempt was made to reveal the new data, or suggest the D.O.E. published data might have been inadequate.

The memorandum described "a scramble" undertaken by D.O.E. to try to water down the import figures. An official contacted eleven of the nineteen U.S. companies that had reported their import figures to I.E.A. This all-out effort to validate the "crisis" resulted in a downward revision of only 0.2 million barrels a day and left D.O.E. with evidence of "a crude and product import surge of almost one million barrels per day for February."

In conclusion, the memorandum fired this broadside at D.O.E.

It is interesting to speculate on the reason behind D.O.E.'s consternation over news that should have been good news to the American consumer—a reported increase in U.S. oil imports at a time of feared scarcity.

One can only assume that this good news was viewed by D.O.E. as a threat to its strategy of exploiting this crisis as an opportunity to drive home to the American people the fragility of our energy situation, the need for conservation, the need for higher prices—the need to sacrifice. Some claim this is Schlesinger's Gulf of Tonkin. The last thing a crisis manager wants is for his crisis to evaporate before his eyes. Again, the American consumer is being sacrificed on the altar of an alleged energy crisis.*

The other industry explanations of the shortage that wasn't were as spurious as the Iranian shortfall. The argument that consumption had increased, helping to cause the "shortage," was the reverse of fact: consumption was really down. I.E.A. reports showed that U.S. oil consumption for the period from January to April 1979 was nearly 1 percent below the level for the same period in 1978. An interesting memo compiled by the staff of Ohio Sen. Howard Metzenbaum, containing figures on travel on the Ohio Turnpike, showed that there had been a 4-percent decrease in the number of cars traveling, a 6.8-percent decrease in miles driven and a 3-percent decrease in miles per trip.

Regarding the oil companies' claim that bad weather in 1979 had hindered oil production, the memorandum added this significant observation: "Weather last year was worse than this year. It included one of the worst blizzards in many years. . . ." (The General Accounting Office, the congressional watchdog, in a report made public in September, would also discredit Big Oil's alibi that bad weather had had anything to do with a reduction of U.S. oil production.)

Turning this noncrisis into a crisis that would cripple the nation for months took nothing short of collusive manipulation among the Big Oil companies and their trained seals in D.O.E. The evidence shows that all stops were pulled out to orchestrate the kind of panic that would drive up prices.

* This "Gulf of Tonkin" report from Dingell's staff was one item that I did not have when I wrote the original *Nation* article. It did not come into my possession until some months later. I have used it here because it was, again, confirmation from a responsible official source of the conclusions I had already drawn from the statistics I have cited.

In writing the *Nation* exposé, I described a couple of these manipulations (for which I was later criticized). The first involved the deliberate stockpiling of oil, both on shore and at sea. Frank Collins, head of the Oil, Chemical and Atomic Energy Workers Union—the union whose men crew the tankers—described some of these maneuvers for me.

Tankers bringing finished products from refineries in the Gulf of Mexico and the Caribbean had been ordered, he said, to steam at only ten knots instead of seventeen, adding days to the voyage. When they arrived at their destinations, they sometimes found storage tanks so full they could not discharge their cargoes, and so many tankers were waiting to disgorge their contents through clogged channels that turnaround time in port had been lengthened from the usual two days to five days. This stockpiling at sea was one method of slowing deliveries and helping to starve gas stations on land. Big Oil threw a conniption over this charge, ridiculing it as preposterous, but I had checked it out with some of Collins's regional officials, and they had backed up what he had said.

Furthermore, virtually the same report had appeared in the pages of *The New York Times* on July 3. Collins had cited this article to me as accurate and typical of what was going on. The *Times* had reported that, in May, the tanker *Mobil Aero* sailed from Beaumont, Tex., with a load of gasoline for three cities in Florida. She had had to return to Texas with 132,000 gallons still in her hold because storage tanks in the Florida port cities were all full. A sister ship, the *Mobil Fuel*, making the same trip from Beaumont to Florida, had had to return to Beaumont on May 8 with 115,000 gallons of super no-lead gasoline that she had been unable to deliver. Significantly, *The Times* article had reported:

And from April 16 to May 10 Mobil's Gulf Coast tankers operated at reduced speed, with ships taking nine days to make a trip from Beaumont to Boston that normally takes five. Tankers that customarily loaded in sixteen hours took up to four days to fill their holds, and the time spent unloading increased from twenty to thirty-six hours.

The second point in the *Nation* article for which Big Oil took me to task involved whether American oil companies, in a time of domestic "crisis," were actually diverting oil from our shores to Europe. I had come across a suspicious trail in the C.I.A. analysis, showing we had exported more oil in this "crisis" year than we had in either 1977 or 1978. The figures showed that exports had ranged from 329,000 barrels daily in January to 455,000 barrels daily in both April and May. Yet in lush 1977, exports had ranged from only 192,000 barrels to 288,000 barrels daily. Energy producers immediately derided this analysis because, under American laws, no oil exports were permitted except in an even exchange with another nation. It was, as the industry called it, "a wash."

Perhaps so, but there *were* strange doings. A lot of Alaskan oil was being shipped through the Panama Canal to refineries in the Caribbean. What happened in the Caribbean was another story. This much, however, is fact: American-owned refineries there, just at this time of our greatest travail, were shipping huge amounts of refined products to the spot market in Rotterdam, where prices had risen to $25 to $30 a barrel.

Energy User News, a trade publication, reported that early in the year Texaco had diverted 65,000 barrels a day from its Caribbean refinery to Rotterdam. Although the action was not illegal, the publication said, "it did affect this country's shortages and thus raise prices." Texaco was not the only offender. Three other Caribbean refineries—the Bahamas Oil Refining Company, partly owned by Standard Oil of California; the Lago Oil and Transport Company, an Exxon subsidiary in Aruba; and Shell Curacao NV—were all sending tankers to Europe. Significantly, Texaco, while shipping tankerloads of fuel to Rotterdam, cut back Boston

Edison Company by 50 percent, or 4 million barrels, in June.

Brian Ross, NBC's crack investigative reporter, later pursued this trail, driving mighty Exxon to the wall. In April, President Carter had suspended the federal Clean Air Act for the entire state of Florida, permitting Florida Power & Light to burn environmentally harmful high-sulphur fuel. The reason: Exxon's Aruba refinery, which had been supplying FP&L with low-sulphur oil, had cut its allotment in half. Ed Hess, an Exxon vice president, told Ross that the refinery had decided to make more gasoline and diesel fuel—and so it couldn't fill FP&L's needs. The changes, Hess said, "were made to meet pressing demands in the United States."

But Ross discovered that many of Exxon's tankers, instead of coming to the United States, were going to Europe. Hess tried to explain that this happened only rarely, but NBC got Lloyd's of London to check—and Lloyd's reported that fifteen Exxon tankers had left Aruba for Europe in the first part of the year, "carrying more than three million barrels of petroleum products."

At this point, Ed Hess disappeared from the picture, and A. K. Wolgast, manager of planning for Exxon International, took up the task of explaining. Shown the Lloyd's study, Wolgast admitted that fifteen tankers had, indeed, strayed to Europe. He wouldn't say where four of these tankers had gone, leading to the inevitable assumption that they had disposed of their cargoes on the high-priced spot market— and to hell with clean air in Florida.

Such was the situation when I labeled the whole 1979 gasoline "crisis" a fraud. The reaction in Carter's oil-dipped administration was almost immediate. I heard the first rumbles through a couple of telephone calls. "Boy, did that article rattle some rafters around here!" one of my congressional sources told me. Even more surprising was a call I got from a girl working in one of the catacombs of D.O.E. "That was a dynamite article!" she told me. She explained that she did not have any contact with the autocratic Dr. Schlesinger herself but that her bosses, who did, had been furious about what was going on. It was certainly nice to know from inside D.O.E. itself that I had been right on target; but it soon became obvious that I had hit some high-and-mighty nerves that were not as happy with my effort. The Carter administration did its damndest to defend the oil companies.

The president, in one of his impersonations of the guardian of the public interest, had issued on May 25 a seemingly stern order to the departments of Energy and Justice.

You shall jointly conduct a comprehensive investigation of the apparent gas shortage situation, using all available and appropriate authority and resources at your disposal, to determine whether there is reason to believe that the apparent shortfall is a result of concerted activity by firms at the refining and/or marketing level, or of excessive stockpiling or hoarding of supplies.

This mighty joint effort produced a whitewash of the oil companies that the White House happily released on August 6, a timing that enabled it to rebut everything I had said in the *Nation* edition then on the stands. The Energy Department carried the ball, submitting a fifty-three-page report to the president that said the oil companies hadn't been at fault; it had found no evidence of hoarding by the refineries. At the worst, "some refiners might have been conservative in their use of [oil] stocks. . . . This pessimism appears to be due in large part to their pessimistic views about the future availability of oil imports." Secretary Schlesinger, who was about to be replaced, confessed that D.O.E. had been at fault through failures in its allocation system. Part of the reason for the long gas lines, he said, was that the department had allocated gasoline to areas "where the cars were not." Unmentioned, of course, was the fact that the industry-consultant-D.O.E. relationship had given the oil companies virtually a free hand in determining allocations.

The *Washington Post* wasn't about to let the administration get away with this flimflam. Two of its investigative reporters, Jonathan Neumann and Patrick Tyler, questioned a number of D.O.E. officials and determined that the supposedly authoritative report presented to Carter was a complete phony. D.O.E. had conducted no investigation, as it had been ordered to do; it had simply presented the president with what was essentially a rehash of policy statements top officials of the department had been making for months. One D.O.E. executive described the performance as "a propaganda exercise."

What had happened, Neumann and Tyler disclosed, was that, when the presidential order came down, some top D.O.E. officials were summoned to the office of Deputy Secretary O'Leary—the official with whom Representative Rosenthal had clashed so bitterly in the spring. On O'Leary's instructions, the preparation of the report was handed over to General Counsel Lynn Coleman, whose links to oil interests have already been described. Coleman and Carlyle Hystad, of D.O.E.'s policy branch, did most of the work, with Hystad relying heavily on speeches he had previously written.

The result: twenty-two of the twenty-seven information sources cited in the appendix of the report were not used in the report; five of these same twenty-two sources had already been stamped "not valid," according to Al Linden, of D.O.E.'s energy information administration. Lynn Coleman rejected suggestions to pursue key issues, such as whether the oil companies had deliberately cut back on production. D.O.E. ignored the presidential order giving it authority to subpoena witnesses; it called no one; it conducted no audits or field investigations to determine, as the president had ordered, whether there was "concerted activity by firms at the refining and/or marketing level." In other words, as Linden admitted, all the information on which the report was based—what there was of it—had been supplied by the oil companies themselves.

Neumann and Tyler wrote, "One official likened the investigation to a man who dropped a dollar bill on a dark street and instead of looking for it where he dropped it, he looked for it under a street light."

This page-one exposé in the prestigious *Washington Post* could hardly have escaped the attention of the Man in the White House. One might think that he would have been furious, righteously incensed at the way his direct order had been flouted, at the way he had been duped into issuing under the White House's august imprimatur a report that was so basically dishonest. But was Jimmy Carter furious? Did he disown the deed? No. Jimmy Carter was as silent as a mummy. Worse, he and his administration spent the next months of his term trying to justify the unjustifiable, trying to exculpate the oil companies—and trying to squelch any conscientious, lower-echelon expert who attempted to point out the truth.

8

Carter Disowns

His Own

Jimmy Carter was shocked, though he had no reason to be. He was in Tokyo conferring with the leaders of six other industrial nations on June 29, 1979, when OPEC dropped its second bombshell in six months.

Throughout 1977 and the first eleven months of 1978, OPEC prices had remained fairly stable. Then, in December, in a meeting in Abu Dhabi, the leaders of OPEC had decided to increase crude oil prices 14.5 percent in 1979. It had been presumed that this decision would fix prices for the year, but on June 28 in Geneva, OPEC voted another hike. This raised the price of a forty-two-gallon barrel of Arabian light crude from $14.55 to $18, another 14-percent boost. Countries such as Nigeria, Libya and Algeria, which produced the best OPEC oil, were not bound by this new ceiling, however, but were allowed to post prices of $23.50 a barrel. And the cartel decreed that any member could impose another $2-a-barrel surcharge if market conditions permitted.

The Western industrial nations, their economies already in disarray from high energy costs, were shocked. *The New York Times* reported:

In Washington, the price increase dismayed the administration and private economists, who predicted that the rises would bring more inflation and less economic growth. A sharp drop in the consumer's purchasing power was also forecast, with a rise of 5 cents a gallon for oil products. Purchases of automobiles, houses and other costly items were expected to be hardest hit.

In Tokyo, President Carter, stern and grim of face, angrily denounced OPEC, declaring that its action would cause worldwide suffering and further damage the economies of the industrial nations. OPEC was painted as the irresponsible villain of a new world energy crisis.

Actually, Carter should have known better, for OPEC had warned the West publicly before the June escalation that the producing states would have to raise their prices unless the United States acted to curb Big Oil machinations.

On June 1, Saudi Arabia's official Riyadh radio had broadcast a blunt warning.

It appears that the OPEC countries are about to enter a war of real confrontation with a number of parties. The oil-producing countries can no longer yield to additional pressures, and they can no longer remain uninvolved while viewing a number of alarming scenes on the world markets.

The operation by certain major powers of huge stockpiles or price manipulation by companies at the expense of the producing countries and at the expense of the countries of the world which are aspiring for more development and prosperity, all these negative manifestations in the relations between the producing countries and the various parties must be resolved in favor of our countries, our peoples and resources.

The broadcast called for a united stand and "the adoption of measures to confront the world companies." It called for "abolition of direct dealings with the oil markets [the spot market]" and new commitment to a unified price. "The industrial states and the world companies have benefitted from the various prices among the OPEC states," the Riyadh broadcast said.

The references to stockpiling and market manipulation certainly indicated that the Arabs were aware of the excessive profits Big Oil was reaping through its various devices: the phony gasoline "crisis" that had driven prices unjustifiably high, the manner in which tankerloads of petroleum products had been diverted to the high spot market in Rotterdam, the way these and other moves had been coordinated to enhance Big Oil's profits at the expense of OPEC.

What the Arabs were saying, in effect, was simply this: *You* are taking *our* oil, manipulating the market with it and making extortionate profits from it. Don't expect us to sit idly by while *you* are making fortunes from *our* oil in which *we* do not share. We are going to get ours, too. One could hardly blame them.

Yet Jimmy Carter throughout the rest of his administration would refuse to recognize that the U.S. multinational oil companies were as responsible as OPEC for the nation's plight. Perhaps more so. Anthony Sampson, the British journalist and expert on the international oil scene, wrote that oil traders in Geneva were telling him that "despite rising OPEC prices, the real villains in the gas shortage were the Western powers—including America—which have steadfastly abdicated their responsibility to control, monitor and indeed force the oil giants to act within the restraints of diplomacy and national interests."

The same message was delivered later to Treasury Secretary G. William Miller when he conferred with Saudi officials in Riyadh. The Saudis warned Miller that yet another round of price increases might be imposed when the OPEC nations met in Caracas, Venezuela, in December unless the administration and Congress acted to cut back on American oil company profits.

"Their message is: Either you put on a windfall profits tax, or we will be raising prices," Miller said. He added, "They feel they've been taken advantage of by the oil companies."

Only Jimmy Carter did not feel that way. The godfather and protector of Big Oil, he returned to Washington to prepare for his Sermon from the Mount. On his desk when he arrived was a June 28 memorandum prepared by Stuart E. Eizenstat, the White House advisor on domestic affairs. Eizenstat laid out the situation in stark terms. He informed the president, "Sporadic violence over gasoline continues to occur. A recent incident in Pennsylvania injured forty."

"The latest C.P.I. [Consumer Price Index] figures," he continued "have demonstrated how substantially energy is affecting inflation—gasoline prices have risen 55 cents since January."

He described the public backlash in these terms:

I do not need to detail for you the political damage we are suffering from all of this. It is perhaps sufficient to say that nothing which has occurred in the administration to date . . . has added so much water to our ship. Nothing else has so frustrated, confused, angered the American people—or so targeted their distress at you personally, as opposed to your advisers, or Congress or outside interests. Mayor [Edward] Koch [of New York] indicated to me (during a meeting the vice president and I had with the New York congressional delegation on their gas problem) he had not witnessed anything comparable to the current emotion in American political life since Vietnam.

Though Eizenstat thought the Vietnam analogy might be "strained in many ways," he nevertheless returned to it in the series of recommendations he made to the president, writing, "You must address the enormous credibility and management problems of D.O.E. which equal in public perception those which State or Defense had during Vietnam."

Carter would never heed this advice. He made some cosmetic changes, replacing Secretary Schlesinger with Charles W. Duncan, Jr., but retaining the same Big Oil favorites in the higher echelons of D.O.E. and doing nothing to independently monitor or question Big Oil data on which department policy would continue to be based.

Repeated proddings in memoranda prepared by trusted aides in his own administration had no effect. On October 29, 1979, Eizenstat and Alfred Kahn, the anti-inflation czar, sent a joint memorandum to the president. It urged Carter to summon about a dozen top oil company executives to the

White House for a "jawboning" session about high petroleum prices. The memorandum made its points in the following terms:

- Increases in the price of crude oil by the OPEC nations can account for much less than one-half of the price increases so far this year in refined petroleum products.
- Increases in the cost of purchased refined products may take that total up *to* one-half, but only one-half.
- Gross margins for refiners increased 6.9 percent during the three months ending in March, 15.7 percent during the three months ending in June, and 37.4 percent during the three months ending in September (these are actual increases, not annual rates).
- In the first nine months of this year, energy prices have increased 33.3 percent (actual, neither seasonally adjusted nor annualized), while gasoline prices have increased 45.7 percent (actual) and heating oil prices 55.6 percent (actual).
- From April through August, energy alone added almost five points to the C.P.I.
- There are indications that some major oil companies tried to cover up the size of their third-quarter profits.

The "jawboning" session never took place. According to the *Washington Post,* Eizenstat, whose name was with Kahn's on the memorandum, suddenly had a change of heart and "ruled out" such a meeting (on whose instructions, one wonders). He opted instead for a meaningless conference with Energy and Treasury department secretaries.

One of Kahn's own aides made one more determined effort. In a memorandum that was quickly squelched officially but soon became widely circulated in Washington, Terence L. O'Rourke drew a devastating picture of the manner in which Big Oil had manipulated both the domestic and international markets. Entitled "A Policy to Fight Inflation in Oil Prices," the memorandum was dated November 7, 1979.

O'Rourke began by describing the intricate manipulations by which Big Oil had put the squeeze on independents, forcing them to go into the high-priced spot market—and how Big Oil had *then* used the higher prices the independents were forced to charge as a cover for justifying Big Oil's own leaping profits. O'Rourke wrote:

The bulk of foreign oil traded in the international markets and imported into the United States is controlled by a handful of major international companies. Other companies buy all or most of their foreign oil from them. In recent months . . . these major companies reduced their third-party sales to other companies in order to meet their own needs and/or divert supplies in order to take advantage of high spot market prices. At the same time, they greatly expanded their mark-ups on remaining third-party sales. Their customers who were cut back were driven into the very thin spot markets for oil where they bid up prices to extraordinary levels. They imported this oil at vastly inflated prices into the United States where it has had the double impact of driving prices up for both domestic crude oil and refined products.

In that paragraph, O'Rourke spelled out the very kind of Big Oil manipulation against which the Saudis had protested and which they had used as a justification for their June increase in crude oil prices. O'Rourke then described the financial benefits to the Big Oil intriguers.

In past months, those few major companies who control the bulk of foreign oil moving in world commerce were able to reap immense profits, because (1) they were assured of adequate supplies, as a result of the control they exercise; (2) they purchased their crude oil supplies at the lowest prices and often resold a portion of it at vastly inflated mark-ups in the limbo of international markets; and (3) they sold their refined products at market prices which reflected the costs of refiners who were buying crude oil at the highest prices.

Such were the details of a scam of colossal proportions, worked on an international scale by some of America's most powerful magnates; its victims were not only the American people, but indeed the people of other nations whose economies were affected by this giant ripoff of inestimable billions of dollars.

O'Rourke made several recommendations, including the suggestion that the government itself might set up a federal authority to purchase and import crude and refined oil products in bulk, thus offering competition in the international market and preventing a few huge companies from determining world prices. He obviously had little hope that such a proposal would meet with favor; and so he recommended more strongly that allocation programs should be amended and more strict control exercised over petroleum price standards.

O'Rourke's memo went into the official discard pile, suffering the same fate as another detailed and powerfully reasoned memorandum that had preceded it. This one had the distinction of having been killed because Jimmy Carter was throwing a tantrum about suggestions that his and Big Oil's precious "energy crisis" didn't exist.

The study in question was prepared in April 1979 by Cathyrn Goddard, director of the U.S. Treasury's Office of International Energy Research, and J. G. Polach, an economist in her office. It dealt with the first quarter of 1979 and the Iranian shortfall myth.

Goddard and Polach found (as had the C.I.A., Customs, the F.T.C., the Washington state attorney general, and Representative Rosenthal) that there had never been a shortfall. To this, they added one additional, stunning conclusion: the Big Oil companies had been actually *better off* without Iranian oil because the substitute oils they got from places like the North Sea, Nigeria, Libya and Algeria were much lighter, sweeter crudes that took less refining and produced more gallons of product per barrel.

During 1978, the five major U.S. oil companies were deriving about 210–220 thousand b/d [barrels per day] of crude from

Iran up to November and December when their liftings declined. They stopped altogether in 1979. By then, four of the five U.S. majors participating in the consortium were lifting and exporting supplemental crude from Saudi Arabia and also the North Sea, Nigeria, Libya, Indonesia, etc. Even on the basis of their shares in ARAMCO (the so-called Seven Sisters cartel) alone, they were more than fully replacing the supplies lost in Iran. Moreover, they were receiving part of Iran's output in November and December, in addition to the increased shares in supplemental liftings they claimed as shareholders in ARAMCO during this period. It is, therefore, reasonable to assume that they entered the first quarter of 1979 *with a considerable crude surplus*, compared to average crude supplies they received in the first half of 1978. [Italics added.]

Goddard and Polach flatly contradicted the Big Oil contention that the replacement crudes were of poorer quality than Iranian oil and, therefore, produced less gasoline. After carefully analyzing the contents of the various crudes, Goddard and Polach concluded that "the quality of available crude supply improved (rather than worsened) in terms of gasoline and middle-distillate yields."

They found that Nigerian crude had a processing advantage of $1.50 a barrel as the result of its lower sulphur content. They found that "substituting Algeria's Saharan blend for Iran's Light will increase the yield of gasoline by 7 percent per barrel of throughput, including an increase in the yield of unleaded gasolines by 2 percent. When Nigerian Light is substituted for Iran's Light, the yield in mid-distillates goes up by 12 percent and that of gasoline by 3.4 percent."

Analyzing the supply and price structure, Goddard and Polach concluded that there was no justification for the ruinous price increases imposed on the American public. "Clearly," they wrote, "they [the majors] were still maximizing their profits." In conclusion, they gave this verdict: "Iran's shortfall in exports has not produced any real shortage of crude in the United States. The substitution of African and North Sea crudes improved the yields in gasolines and mid-distillates, [and] actually decreases the crude requirements. . . ."

This draft memorandum was widely circulated in the Treasury and State departments; and when its impact was understood, the royal stuff hit the fan. An interoffice memo to Deputy Assistant Secretary John Karlik, Goddard's boss, described the almost terrified reaction of then–Secretary of the Treasury W. Michael Blumenthal.

Somehow or other Mike heard about the Goddard paper and is disturbed about possible leaks. Apparently there was adverse comment at yesterday's cabinet meeting about a Brock Adams statement on energy supplies, and Mike doesn't want anything attributed to Treasury staff that undercuts the official position. Please make every effort to keep the paper under tight control until its [sic] been reviewed internally.

That memo effectively clamped the lid on Goddard's research. Acting Deputy Assistant Secretary Charles Schotta, who had the reputation of favoring Big Oil anyway, wrote a memo discounting Goddard's findings and imputing to her shoddy research. Schotta added that Goddard's report was "quite likely to be confusing to those in the public and in Congress who, even now, doubt that there is an energy crisis facing us." Schotta pledged his services in "minimizing any adverse effect which could stem from this incident."

Cathryn Goddard, the record shows, was courageous and committed enough to try to buck this cover-up. On May 18, she fired off a memorandum to her boss, John Karlik. The first paragraph read:

In the light of Charles Schotta's unsupported conclusions as to the analytic validity of Jay Polach and my report, I recommend Assistant Secretary [Dr. C. Fred] Bergsten be given a copy of our text to allow him to draw his own conclusions.

Cathryn Goddard thus committed the unforgivable heresy in a bureaucracy. She had gone over the head of a higher-up. After that, she became a pariah in the Treasury Department. Her report was deep-sixed without explanation. She was by-passed and left to twiddle her thumbs. After she had had as much of this as she could take, she resigned in disgust.

When I talked to her some months later at the office of a consulting firm in Alexandria, Va., where she had found a happy haven, she described the whole distressing experience. She explained that her report had been a draft, as it had indeed been plainly labeled, and that she had not been fighting to get it accepted as final. All she had wanted was to get it seriously considered. "It should have been taken up and weighed on higher levels, but it never was," she said. "How are you going to make intelligent policy if you don't take up and consider all the facts? But that just didn't happen."

And so Cathryn Goddard joined the long line, stretching back to John Sawhill and Joe McNeff and Brock Adams himself, of those who had had their heads handed to them or had been forced out of government service for crossing Big Oil and Jimmy Carter. It all reminded me of Leo Durocher's famous saying "Nice guys finish last." There has to be a corollary to that about the characters of those who finish at the top, but I'm afraid it wouldn't be gentlemanly, however well deserved, to apply the biological term that comes to mind to the distinguished hierarchy of the Carter administration.*

---

* President Carter, having taken office with the pledge "I will never lie to you," failed to heed Stuart Eizenstat's warning that his credibility was at stake in the 1979 gasoline "crisis." When the first D.O.E. report exonerating the oil companies was exposed as a phony, Press Secretary Jody Powell excused it as being "premature"—not patently dishonest. Almost a year passed. Then, on July 17, 1980, the White House issued studies by the Justice Department and D.O.E.

that held the oil industry as pure as Snow White. The Justice
Department study—"Report of the Department of Justice to the
President Concerning the Gasoline Shortage of 1979," signed by
Attorney General Benjamin R. Civiletti—found no evidence of collusion, price-fixing or antitrust violations. It held that the 1979
shortage had resulted from the Iranian shortfall, a drop in domestic
production caused by bad weather, and D.O.E.'s fouled up allocations
of supplies (ignoring the fact, of course, that these allocations were
based on industry-supplied data). A close examination of the report—
a feat that was not performed by newspapers like *The New York
Times*, which blandly heralded it with a page-one headline reading
U.S. OIL COMPANIES HELD BLAMELESS IN '79 GAS SHORTAGE—reveals
that this belated effort was as flawed as D.O.E.'s original whitewash.
The Justice report clearly states that it is based on information
provided by D.O.E. and the oil companies, both sources hardly liable
to convict themselves. It arrived at an "Iranian shortfall" by using
figures for a nine-month period, ignoring the fact that the gasoline
lines would have had to have been caused by shortages in the first
three months. Furthermore, Justice's "shortfall" represented not what
happened in the real world but *a shortfall of predictions*. On page 36
of the report, Justice explained its methodology. A forecast of supplies
for the coming year is prepared each October by the Independent
Petroleum Association of America (IPAA), described by Justice as
representing "independent oil and gas producers." The Justice report
added that "the predictions prepared by IPAA in October of 1978 for
1979 provide an excellent basis for the Department's analysis." On
page 39, a Justice table compares actual gasoline supplies with IPAA's
"prediction." Even this shows that the critical first quarter of 1979
exceeded the "prediction" by 20,000 barrels a day. Supplies fell behind
the "prediction" in the second and third quarters, leaving an overall
"shortfall" of 463,000 barrels per day. What we are left with, obviously, is a "prediction shortfall" rather than an actual shortfall. One
has to wonder, too, how Justice determined that IPAA was composed
of "independent" experts. IPAA's twenty-two member committee,
listed at the back of the report, was composed of representatives that
included Big Oil's bellwether bank, Chase Manhattan; Cities Service;
Phillips Petroleum; Union Oil; Texas Eastern Transmission; Getty
Oil; Tenneco; Standard Oil of Ohio; Marathon Oil; the Hughes Tool
Company; Conoco; and Ashland Petroleum. Quite an independent
group! One can only conclude that this second whitewash of Big Oil
was as phony as the first, but it enabled aggressive Mobil to proclaim
in advertisements and on television that the oil industry had been
"Cleared!" citing as evidence Benjamin Civiletti's report.

The Wages of

9

Chicanery

Big Oil manipulation of the market in 1979—the creation of the phony oil crisis, the diverting of tankerloads of refined products to the spot market, the deliberate driving up of prices—made the year the most profitable in the history of the industry. The wages of chicanery kept Big Oil cash registers jingling with incredible multibillion-dollar profits. The American people had certainly suffered—and were to continue to suffer—but not Big Oil.

Reports of profits so huge that Mayor Ed Koch of New York called them "an obscenity" began to flow in during late October 1979, when the oil companies revealed the results of their third-quarter operations. Some of them like Exxon were extremely bashful about it.

The Associated Press noted that "you'd have to read three pages into the firm's press release on the quarter's results" to find out that Exxon's profits had surged 118 percent. Its net profit for this single quarter totaled *$1.145 billion.* Exxon, as the largest of Big Oil, chalked up the greatest one-quarter profit; but it was by no means alone in surpassing previous records, nor was its increase the largest in percentage terms.

Texaco's profit for the third quarter leaped 211 percent. Others among the biggies reported these percentage increases: Standard Oil of Ohio, 191; Conoco, 134; Mobil, 131; Gulf, 97; Sun Oil, 65; Cities Service, 64; Phillips, 62; Marathon, 58. Overall third-quarter profits for the eighteen largest oil companies totaled *$5.69 billion, compared to $3.04 billion in 1978.*

"The big surge in third-quarter earnings is causing an embarrassment to the oil companies," said Sanford Margoshes, the oil industry expert for the Wall Street firm of Shearson Hayden Stone.

Market experts noted that the profit surge had begun in the first quarter and had continued longer than had been expected. What they could not foresee was that the profits of 1979 were like the first-stage booster of a rocket on take-off; this was only the beginning of the flight into the stratosphere.

The companies had all kinds of explanations. The bulk of their profits, they said, had come from overseas operations (the spot market, for one?) and not from domestic refining and pricing. They pictured themselves as almost struggling to make a profit on their domestic operations, ignoring the obvious fact that huge increases in refinery margins and prices, as cited in even the Carter administration's internal reports, must have contributed hundreds of millions of dollars reamed out of the pockets of American consumers.

Another justification, one especially favored by Texaco, held that profits in this shocking multibillion-dollar range were essential to give the majors the cash needed to explore for more oil and gas for the benefit of us all. Texaco, for example, said that it had budgeted $10 billion for exploration and capital spending in the next five years.

This explanation had a kind of superficial logic—that is, it did until one looked at what the oil companies were doing with their surplus cash. On September 9, a month before the third-quarter reports began to come out, Exxon spent $1.17 billion to acquire Reliance Electric Corporation in Cleveland. This marked one of the largest cash takeovers in the history of American capitalism up to that time, and it seemed to have no connection with Exxon's major responsibility of producing more oil and gas for the needy American economy.

Reliance was not drilling for oil or gas in Cleveland; it was making electric motors. What connection could this have with Exxon's business? Well, Exxon explained, it expected through Reliance to be able to develop a new type of electric motor that would save enormous amounts of energy. So its investment was really related to the energy problem, you see. A lot of skeptics didn't see. Other electric firms had tried to develop such supersaver motors without success; the idea seemed something like the fantasy of the perpetual motion machine. Exxon's experience with Reliance was no different. It failed, as others had, to develop the miracle motor, and by the time a couple of years had passed, it had to confess defeat.

Exxon was not the only Big Oil company to squander its billions in business adventures that had nothing to do with relieving the anguish on the gasoline lines. Even before the big bucks of 1979 began to clog Big Oil treasuries, industry profits had been so lush, thanks to the benefits derived from the 1973–74 Arab oil embargo, that Big Oil firms had been taking over vast sections of industry, both here and abroad.

In 1976, Mobil spent $1.8 billion to acquire Montgomery Ward, expecting to strike oil by drilling through the floors of the huge merchandising chain, no doubt. (Even this huge expenditure was only the beginning. By the end of 1981, Mobil had poured another $575 million in interest-free loans into Montgomery Ward in the effort to keep the one-time rival of Sears Roebuck solvent.)

In that same year of 1976, Atlantic Richfield (Arco) had spent $784 million to get control of Anaconda, the second largest copper company in the nation; in that year, for some reason unfathomable to energy experts, Amoco had purchased a London newspaper, *The Observer*.

The asininity of some of these financial fliers might have been cause for sardonic laughter had there not been a more ominous aspect to the invasion of broad sectors of the American economy by Big Oil's billions. By 1979, oil companies had acquired two of the four largest coal companies in the nation, fourteen of the twenty largest reserve coal fields and two of the three largest uranium producers.

The extent of such takeovers and the potential dangers they posed were spelled out in a letter that Energy Commissioner Jacobson of New Jersey wrote on May 9, 1979, to Sen. William Bradley. Jacobson's list covered the period 1974–78, and it is long.

Company	Acquisition	Product or Service
EXXON	Compania Minera Disputada (Chile)	Copper mining
	Jefferson Chemical (Great Britain)	Veterinary chemicals
TEXACO	Jefferson Chemical (Canada)	Chemical marketing
GULF	Kewanee Industries	Chemicals
STANDARD OIL OF INDIANA (AMOCO)	Analog Devices	Components
	Cetus	Microbiology
	Sinclair-Koppers	Plastics, dyes
	The Observer	
ATLANTIC RICHFIELD (ARCO)	Anaconda	Copper mining
	Continental Cables & Conduits	Electrical conduits
	I.C. Engineering	Process control
	Solar Technology International	Photovoltaic cells
SHELL	Polymer Division of Witco Chemical	Plastics
	Starla-Werken (Sweden)	Automotive exhaust systems
	ETS R. Bellangr (France)	Automotive exhaust systems
	Harmo Industries (Great Britain)	Automotive parts

Company	Acquisition	Product or Service
TENNECO	International Foam Division of Holiday Inns	Flexible polyurethane foam
	L.D. Properties	Almond orchards
	Monroe Auto Equipment	Hydraulic shock absorbers
	Philadelphia Life Insurance	Insurance
	H.P. International	Industrial distribution
	St. Johnsbury Trucking	Trucking
	Audio Magnetics	Tape cassettes
SUN	Stop-N-Go	Retail grocery chain
	Walter Norris	Industrial distribution
	Applied Financial Systems	Computer software
	Kar Products	Equipment distribution
UNION OIL	Molycorp	Rare earths
	Adtek	Design, construction and engineering
OCCIDENTAL PETROLEUM	Squamish Chemicals (Canada)	Chemicals
	Zoecon	Pesticides
	Anchor Construction	Heavy construction
	McClinton Brothers	Construction materials
	North Western Arkansas Asphalt	Heavy construction
	Lehigh Valley Chemicals	Chemicals

Company	Acquisition	Product or Service
ASHLAND	Levingston Shipbuilding	Shipbuilding
	AB&H Processing	Mining
	Coastal Chemicals	Chemicals
	Commonwealth Equipment	Mining
	General Oils	Petroleum
	Highland Tractor Service	Mining equipment

Jacobson wound up his letter to Bradley by asking, "Isn't it peculiar that the two most profitable periods in the entire history of the oil industry were, first, 1974—when everyone was waiting in those long lines for gasoline—and second, the first quarter of 1979 [the letter was written before those third-quarter reports came out], when we're told the shortfall of crude from Iran has caused a 'shortage.' "

All of this may seem incredible. It is so simple to blame OPEC for our energy problems. When OPEC embargoed shipments of oil in 1973–74, when OPEC crude oil prices jumped twice in six months in 1978–79, it couldn't have been Big Oil's fault, could it? Prices had to go up, didn't they? Didn't that hurt Big Oil as much as the average American?

The answers are that of course OPEC bears a large share of the responsibility; but, at the same time, OPEC is just about the best thing that ever happened to Big Oil. Every OPEC price boost must have had the emperors of Big Oil jumping in the air and clicking their heels for joy in the exclusiveness of their executive suites, for every OPEC increase translated into hundreds of millions of dollars of extra profits for them. Consider: The oil companies normally have between 330 and 350 million barrels of crude oil in inventory. This is oil purchased and transported across the seas months before an OPEC price increase. Thus, OPEC's minimum June 1979 hike from $14.55 to $18 a barrel meant a $3.45-a-barrel inventory windfall for Big Oil, since those millions of barrels in inventory had been purchased at the price prevailing months earlier. With the new price increase, that oil immediately became valued at the new OPEC benchmark. In other words, if one multiplies 330 million (a minimum figure) by $3.45 (also a minimum since much OPEC oil was priced higher and the spot market operations were increasing prices still more), one comes out with an instantaneous windfall of more than $1.1 billion.

It is not hard to understand how Big Oil chalked up what Mayor Koch called "obscene" profits in the third quarter of 1979. The manipulations that had sent tankerloads of petroleum products to the overseas spot market, eventually driving prices up to between $40 and $50 a barrel, established a price platform vastly higher than OPEC's; and Big Oil, which had bought at the lowest possible prices under earlier long-term contracts, could inflate even more outrageously the value of its multibillion-barrel inventory. At this point, the potential magnitude of the ripoff begins to boggle the mind.

Events of the period show that Big Oil took advantage of almost any pretext to jack up those inventory values at the expense of a suffering American people. Take, for example, an incident that happened on Tuesday, November 13. 1979. Iran had started to resume oil production, but President Carter had just decreed that we wouldn't buy any more oil from Iran.

I was sitting in my favorite luncheonette—a place I frequent because it is an ideal sounding board for what goes on in American life—when my fuel oil dealer friend, whom I have quoted before, came storming in. He was wild with indignation. "Do you know what they've done now?" he roared in a voice that carried throughout the place. ("They" were the oil companies.)

"Do you know what they did?" my friend roared on.

"They've all increased their heating oil prices three cents a gallon. Within twenty-four hours!"

Cautious news stories following the president's announcement of an Iranian boycott had predicted that the cutoff would not cause any immediate shortage and that heating oil prices would probably edge only slightly higher over a period of several months. But the big oil companies couldn't wait. They imposed their new gouge immediately.

The day after my oil dealer friend had exploded in frustration, I found from the touch on my pocketbook just how correct he had been. My own fuel oil deliveryman came to top off my tank, and I found that the price had indeed leaped 3 cents a gallon. When I asked about it, my deliveryman made a face and said, "They all raised their prices Monday night." My price was now suddenly 85.7 cents a gallon, and the mere 92.4 gallons my tank took cost $79.37.

The point of all this is that the oil companies, especially early in the heating season, have a middle-distillate stockpile of between 220 and 245 million forty-two–gallon barrels, just as with crude oil inventories. Thus, an overnight 3-cents-a-gallon boost, completely unrelated to the cost of those billions of gallons of heating oil in inventory, figures out to an oil industry windfall in the nature of $300 million. Overnight. For nothing.

But, the credulous will protest, don't we have a free-enterprise economy? If so, such gouges could not possibly happen because, working with margins of this magnitude, certainly some industrious free-enterpriser would cut prices, underselling the competition and gathering for himself a larger share of the market. Ronald Reagan, living in a past that never was, would doubtless say that this is exactly the way the system would work. The answer is: It never did.

It didn't because the energy industry of the nation is in the hands of a Big Oil cartel. It is a cartel that has the power to drive independent competitors out of business, that has the power to rig the international oil market for its own benefit, that sets and maintains price levels that apply across the board to every company, every dealer.

Example: In May 1979, attorneys general for California and five other western states (known as the Pad V region) went into federal court and lodged an antitrust action charging that ten companies—Standard Oil of California (Chevron), Texaco, Union, Atlantic Richfield (Arco), Exxon, Getty, Gulf, Mobil, Phillips and Shell—had conspired to control the retail gasoline market and to fix prices.

The suit alleges that much of the anguish caused by the 1973–74 Arab embargo was in reality the work of the major oil companies. They had begun in 1972 to cut back deliveries to dealers, driving many out of business. In February 1973, long before the OPEC boycott, the president of Arco was warning of a coming "fuel crunch." Refinery utilization dropped as low as 85 percent in the second half of 1972 (the same tactic that helped create the "gasoline crisis" of 1979), and "voluntary allocations" were imposed by the oil companies to tighten supplies of gasoline delivered to service stations. When the Arab boycott came along, it camouflaged what the companies had been doing and supplied them with an ideal whipping boy.

The Pad V suit, which is expected to drag on through the courts for years, is based in part on information furnished by a former oil company executive, but also on an exhaustive computer analysis of price fluctuations. According to the informant, price-fixing was handled by a secret control group operated by Chevron. This control group functioned, investigators say, in a headquarters in San Francisco, and all oil companies servicing the Pacific area states adhered to the price schedules the Chevron group imposed.

Brian Ross, of NBC, describing the computer comparisons made by the states, said, "In places where the secret group was active, from Tucson to Los Angeles to Seattle, . . . in cities hundreds of miles apart, . . . the average weekly price of gasoline was always the same, and the computer charts show that when prices went up for one oil company, . . . they went up the same for all companies at the same time, . . . station after station."

Assistant Attorney General Tom Boeder of California told Ross, "We start out with Union, Texaco and Standard . . . and we overlay Mobil, Phillips and Shell, . . . and then we add Exxon, Chevron and Arco." The prices were always identical, Boeder said. "These facts," he added, "indicate elimination of competition which is so extreme that there is no longer any competition in the petroleum industry on the West Coast."

If more is needed to illustrate the cartel control of the nation's most basic, most vital business—energy—one has only to look at the international conspiracy that controlled the production of uranium and dictated its price. Throughout the 1970s, one national administration after another held up the mirage of nuclear energy as a cheap and virtually inexhaustible source of electric power. The vision was to prove illusory for many reasons, the first being the rapid escalation of uranium ore prices which helped to make nuclear power anything but cheap.

Gulf Oil played a major role in the international cartel, organized in 1972, that drove the price of uranium up from $6 a pound to $40 in 1978. Gulf was the linchpin in a nine-corporation consortium that controlled the uranium market with the help of Canadian and British interests. When the Justice Department began to investigate the cartel in 1976, this multinational aspect of the conspiracy erected sometimes impenetrable barriers for investigators.

Justice Department records in Washington indicate that Gulf on several occasions tried to prevent documentary evidence linking it to the cartel from falling into government hands. A sworn deposition from a Gulf executive, records in several civil cases and a Justice Department memorandum all indicate that Gulf adopted various strategies to keep its cartel connection secret.

According to these records, Gulf instructed key executives not to bring vital documents into the United States; it apparently shipped some documents to offices in Canada, and it filed others with its attorneys in Pittsburgh. The Justice Department was balked when it tried to obtain records hidden in Canada and protected by Canadian law, and Gulf contended that anything filed with its attorneys was immune to government search, thanks to client-attorney confidentiality.

Investigating attorneys recommended unanimously that the nine corporations involved in the conspiracy be indicted for felony antitrust violations. If convicted, the nine corporations could have been fined up to $1 million apiece and responsible officials could have been sentenced to up to three years in prison and fined as much as $300,000. So what happened? Virtually nothing.

The Carter administration washed out the investigation. Just one company, Gulf Oil, was allowed to plead guilty to a minor misdemeanor count and was fined $40,000. It was an action that provoked a lot of criticism.

S. David Freeman, chairman of the Tennessee Valley Authority (TVA), which has to buy millions of dollars worth of uranium for its nuclear plant, told Brian Ross, "All attention is focused on OPEC and the OPEC cartel, but there are a lot of blue-eyed Arabs in this world too who are getting together and following the same monopoly trend."

Senator Metzenbaum was outraged. "A forty-thousand dollar fine to Gulf Oil is tantamount to no fine at all," he said. "Gulf walked away smiling from the courtroom."

Metzenbaum, chairman of the antitrust subcommittee of the Senate Judiciary Committee, called a hearing and confronted John H. Shenefield, the acting associate attorney general who had jurisdiction over antitrust matters. Questioned sharply by Metzenbaum, Shenefield denied that he had been pressured to overrule the investigating attorneys in the uranium case, but he admitted that he had been warned by the State Department that the matter had foreign implications because the governments of Canada and Great Britain were involved along with the multinational corporations.

Metzenbaum indicated that he was highly displeased with the Justice Department's antitrust performance, and he reminded Shenefield of promises he had made in 1977 and 1978 that the antitrust division was about to bring a "shared monopoly case." Shenefield replied that this had been "an inappropriate thing" for him to have said.

During a five-part NBC series on energy issues, Brian Ross questioned Shenefield, pressing him to describe just what the Justice Department had done in the antitrust area. Shenefield recalled "one of the clear success stories of antitrust."

ROSS: What year was that?
SHENEFIELD: That was 1911.

Shenefield was referring to the historic antitrust action taken by an entirely different Justice Department in the era of Theodore Roosevelt. This was the action that resulted in the break-up, for the time being at least, of John D. Rockefeller's Standard Oil monopoly.

What is at stake for the country in the present age of cartels was perhaps best summed up by TVA Chairman Freeman when he told Ross, "The question is, Are we going to wake up in 1990 and find [that] a half a dozen or so major oil companies own all of this nation's energy resources and are charging Arab-style prices for these resources? If we do we will find that they own America."

10

The Merciless

Gouge

Anna Dolan was a small, spry woman of eighty-four who walked with a cane. She ate breakfast at a neighborhood cafeteria in Washington, D.C., and spent many afternoons playing bridge at the New York Avenue Presbyterian Church. She lived with a number of other elderly, lonely persons in the Wisteria Mansion Apartments at 1101 L Street, NW—and there, on Friday afternoon, January 15, 1981, neighbors found her semiconscious in her bed.

She was taken to Howard University Hospital, and there she died. She was a victim of hypothermia.

No one knows—no one in the United States has made any determined effort to find out—how many aged Anna Dolans die every year from hypothermia, a deadly affliction that comes from the sudden lowering of body temperature.

The eight-story Wisteria Mansion (so named for the vines that used to climb its walls and flower beautifully in the spring) is home for the elderly who have no other home. The building's operators had been trying to save on the use of expensive heating oil. At ten o'clock every night, the heat was cut off; it didn't come on again until six o'clock the next morning. In the intervening eight hours, the rooms in the Wisteria became frigid—too frigid for Anna Dolan, whose heart was weak. Normal body temperature is 98.6 degrees, but when Anna was taken to Howard University Hospital, her temperature was recorded at 94. A drop of 4.6 degrees may not seem like much—but it is enough to kill.

Washington, D.C., police noted that Anna's was the second death to be attributed to hypothermia during 1981's brutal winter. (A third death was to be blamed on the same cause the following week.)

Were these deaths rare and unusual occurrences? The best evidence indicates that they were not.

Back in 1979, energy commissioners and congressional energy experts in the Northeast, where heating oil was the principal fuel, had told me they were extremely worried that the high price of No. 2 oil would force many elderly persons to risk their lives by keeping temperatures in their homes too low in an effort to save on their fuel bills. Hypothermia strikes suddenly in such situations, before the victim is aware of danger. As Ruth Wright, whose apartment was next to Anna Dolan's in the Wisteria, put it, "When you get old, you don't even know you're getting cold."

Energy experts to whom I talked felt that many deaths, perhaps thousands each year, were being caused by hypothermia but were being attributed in many cases to heart failure or other natural causes. An assistant in the office of Rep. Toby Moffett cited British studies, the best that have been made. These showed that thousands there die annually from hypothermia and that 10 percent of the aged are at risk. In the United States, this means that some 2.5 million elderly are potential victims.

The Wall Street Journal, in an article in March of 1979 headlined COST OF KEEPING WARM POSES A REAL HARDSHIP TO MILLIONS OF THE POOR, noted:

For the elderly, letting the temperature in the house drop into even the low sixties can be deadly.

That's because many older people are susceptible to "accidental hypothermia," a rapid drop in body temperature that can be fatal. While firm statistics don't exist, there are estimates that thousands die each year from the condition.

A few individual tragedies got brief notices in the press. In Somerville, N.J., Mrs. Alice Condit, seventy, lived alone on a small income and tried to economize on expensive heating oil purchases. Though her family urged her to keep her home warmer, she was determined to keep the temperature low and not buy any more oil until she absolutely had to. She miscalculated. Her grandson, Steven Condit, found her lying on the living-room floor near a couch. She was dead.

Lt. Thomas Stabile, who investigated, said later that Mrs. Condit's fuel tank was empty. "We theorize she fell asleep and then the temperature, already low, fell even lower," he said. "When we found her, the room temperature was only 10 degrees. She just froze to death."

Similar tragedies occurred in Maryland. In one acrimonious House debate over a plan for the government to contribute to fuel oil purchases by the poor, Rep. Parren Mitchell shouted at opponents of the measure, "Dammit, some people in my district froze to death last winter."

All the time, the oil companies, heedless of the suffering and deaths they were causing, drove ever higher the price of No. 2 heating oil. And columnist Mary McGrory wrote that members of Congress "seemed resigned to the idea that Big Oil has the country by the throat."

There were sporadic attempts to find out what was happening, what prices were justified and what were not. D.O.E. itself made one of the first such efforts—then walked away from it.

When fuel oil was decontrolled in 1976, Congress had been given certain assurances: Prices would rise no more than one cent a gallon; they would be constantly monitored; if they seemed to be getting out of hand, remedial action would be taken—even, possibly, to the extent of reimposing controls.

The monitoring system in 1976–77 had been an abysmal failure, as Representative Howard and I had discovered, and so there was pressure on D.O.E. to fulfill its pledges. The

Hearings and Appeals Division was created, and in August 1978 Hearing Officer Melvin Goldstein conducted a broad-based investigation.

Goldstein sought evidence from both consumer groups and industry. The American Petroleum Institute and Atlantic Richfield (Arco) agreed originally to present the industry position, but then both withdrew and refused to appear. Arco explained that it was willing to let its position be represented by the Office of Fuels Regulation of D.O.E., an attitude that seemed to express Arco's confidence that D.O.E. would prove a friendly partisan.

As a result, Goldstein heard testimony from the antitrust division of the Department of Justice; D.O.E.'s Office of Fuels Regulation; the National Oil Jobbers Council; and the Energy Policy Task Force (EPTF) of the Consumers Federation of America. He later commented in his report that the withdrawal of A.P.I. and Arco limited evidence "regarding competition at the refining level of the industry" and affected conclusions as to "whether competition is adequate at the refining level." Goldstein found A.P.I.'s and Arco's reasons for withdrawing "fallacious." He wrote, "The allegations and evidence that petroleum refiners do not engage in active competition have not been rebutted." And he ruled that "the burden [of proof] should now be on the industry."

Using D.O.E.'s own figures, Goldstein found that, instead of prices rising only the promised one cent a gallon, they had jumped 6.4 cents during the 1976–77 heating season, the first under decontrol. By the time the hearing was held in August 1978, D.O.E.'s figures showed that heating oil prices had risen 23 percent, compared to a 12.5 percent rise in the Consumer Price Index (CPI).

D.O.E.'s statistics further showed, Goldstein ruled, that refining charges could not be justified by costs. He found a 23 percent gap between refiners' costs and what they were charging. He noted that "the record does not contain any evidence that the refiners of home heating oil do engage in competitive behavior," and he again intimated that, if evidence of competition existed, A.P.I. and Arco should have presented it.

The Energy Policy Task Force, on the other hand, had analyzed available figures and concluded that the refinery ripoff had amounted to at least $193 million. Using the existing price trends to predict the future, it estimated that the overall ripoff from decontrol in June of 1976 through April 1979 would amount to $692 million in overcharges.

In his decision, Goldstein held that refiners had indeed been overcharging, that the situation was likely to continue and that overcharges from the beginning of control to the end of the 1978–79 heating season would be approximately $331 million.

The Justice Department had contended at the hearings that the only remedy lay in antitrust action, but Justice, the reluctant dragon, gave no indication it was prepared to take such action. On the contrary, its oil-company partisanship seemed clearly indicated by its opposition to even the continuation of price monitoring.

Goldstein discarded the antitrust suggestion, noting correctly that such a case, even if undertaken, would drag on for years through the courts. "It would appear that price controls would provide the most desirable response to the situation," he wrote, thus committing the ultimate heresy. Without controls, Goldstein predicted, "prices will continue to rise at a greater rate than costs in the future." Refiners, he wrote, had already charged consumers "hundreds of millions of dollars more for No. 2 heating oil than they would have been permitted to charge under a price control program," and he added that "there is no workably competitive environment to constrain refiners in levying unreasonable prices for No. 2 heating oil in the future."

These recommendations horrified D.O.E. The hierarchy of the department shelved Goldstein's report and announced that the department would conduct its own study. When this exercise was completed in March 1979, it parroted the oil companies' line, arguing that the companies had "incurred more increased costs than increased revenues." D.O.E. explained that this was due to the "seasonability" of the heating oil business. The unfortunate oil companies had to stockpile No. 2 oil in the summer to meet winter's demands and so incurred "banked costs" before they could sell their product. The asininity of this rationalization was obvious. In the first place, the heating oil business had always been "seasonal" and oil companies had never imposed such high costs, banked or otherwise. Furthermore, throughout the 1978–79 heating season, prices had continued to rise at an ever more rapid rate, just as Goldstein had predicted. And, in the summer of 1979, long after the heating season had ended, prices kept jumping at the astronomical pace of one cent a week.

A record of my own fuel bills, which I have kept, is typical of what happened because, free enterprise being nonexistent, prices everywhere marched pretty much in lockstep. At the end of December 1978, fuel oil was priced at 53.5 cents a gallon, a figure that horrified me at the time. But by March 1979, the price was 59.9; by October 22, it was 82.9; by December 6, 85.9; by December 27, 88.9. In other words, heating oil prices had increased 35.4 cents a gallon in just one year, starting from a base of 53.5—and they were still soaring.

When I first became incensed at the unjustified equality between heating oil and gasoline prices, the industry excuse had been that D.O.E. had permitted oil companies to "tilt" additional costs to heating oil because they hadn't been able to get gasoline up to even the permitted ceiling prices. But after the phony gasoline crisis of 1979, an obliging federal government approved a second "tilt"—this time on gasoline prices. This made James E. Lee, president of Gulf Oil, very happy. In a speech before the Oil Analysts Group of New

York on June 18, 1979, he said, "Beginning in March . . . the D.O.E. has granted belated regulatory relief called 'tilt,' which enables us to recover increased refinery costs at the gasoline pump. Taken alone, this would mean an additional three and a half to four cents a gallon on Gulf's gasoline and would provide us with approximately $250 million a year in additional pretax income."

It is not hard to see why Lee was happy. An extra $250 million a year should be enough to raise the spirits of any man. But tilt number two on gasoline, it seemed to me, should have removed the necessity for tilt number one on home heating oil. Logic said that, since there was now a tilt on gasoline, there was no reason for continuing the heating oil tilt that had been imposed because gasoline couldn't be tilted. Logic said heating oil prices should have come down; instead, they went galloping ahead faster than ever.

Various congressional committees pecked away at this outrage, but they were frustrated, just as Mary McGrory had written, because Big Oil had the nation by the throat.

On September 5, 1979, members of the subcommittee on energy and power of the Interstate and Foreign Commerce Committee of the House questioned Deputy Secretary John O'Leary of D.O.E. The deputy secretary, whose record indicated he was an apologist for Big Oil, testified that D.O.E. was "monitoring refiner, wholesale and residential heating oil prices." He acknowledged that fuel oil in mid-August was being priced at 80 cents a gallon, but ignored the fact that it was continuing to climb at a 1-cent-a-week pace; he argued that "adequate supplies will keep heating oil prices at competitive levels." Naturally, D.O.E. was "very concerned that prices will be much higher than last year and that they may impact severely on the poor."

Then, in the most chilling line in his testimony, he added, "There may be people in our country who will be faced this winter with a cruel choice between food or heat."

It was a flat statement with no trace of human sympathy or concern in it—and no recognition that the choice for many might be between living and dying.

Under questioning, O'Leary admitted that D.O.E. had "strong though not conclusive evidence" that a major cause of rising heating oil prices might be found in refinery profit margins. But he insisted he didn't have enough evidence to call the refinery profits "price gouging."

Others were not so hesitant. James L. Feldesman, attorney for the Consumers Energy Council of America, testified before D.O.E.'s Economic Regulatory Administration on September 26. He upbraided D.O.E. for pigeonholing Melvin Goldstein's report, he charged that D.O.E. had "whitewashed the incredible increase in price since the beginning of this year,'" and he estimated that the heating oil ripoff had cost American homeowners $2.4 *billion* since decontrol.

Three other assessments of heating oil prices virtually destroyed whatever credibility D.O.E. and Big Oil had left. The first was a study made by the General Accounting Office (G.A.O.) at the request of Sen. John W. Warner. Senator Warner, concerned about the way heating oil prices were jumping, asked G.A.O. in June 1979 to assess the validity of the prices and the outlook for the 1979–80 season.

G.A.O. flatly contradicted the testimony O'Leary had given Congress. It found that D.O.E. had stopped monitoring heating oil prices after the 1977–78 season. Since then, it had made no effort to determine "whether heating oil prices are equitable." In fact, D.O.E. contended it had no responsibility to determine whether there was competition among refiners. "We disagree," G.A.O. said. It found that "D.O.E. is not on top of the situation" and that its attitude was "particularly distressing in view of studies which describe the severe economic hardships higher energy bills pose for the poor and the elderly."

G.A.O. put these hardships in dollar terms:

In 1978 alone [before heating oil prices really took off], rising energy costs in the United States caused low-income households to suffer a loss in purchasing power of more than $4 billion over and above that which they would have suffered if energy costs had risen at the rate of inflation. This loss will be even greater in 1979.

That paragraph described a major cause of the recession that was then beginning to engulf the nation, for not only were the poor suffering a loss of $4 billion in purchasing power, but the middle class, whose purchasing power largely sustained the economy, was being bled of uncounted additional billions.

The G.A.O. study, like Goldstein's, found the refiners guilty. Most price increases, the study said, "are due to increased prices charged by refiners," and though part of these could be attributed to higher crude oil prices, "refiners have also increased their gross margins." G.A.O. cited a study by the Congressional Research Service that showed refiners' crude oil costs had increased 14.8 cents a gallon (47 percent) between January and August of 1979, but their selling prices "had increased 25 to 26 cents a gallon (61 percent) during the same period."

In a hearing held on this report, Representative Moffett noted that he had been informed in February 1980 that heating oil prices in Connecticut had risen to a dollar a gallon (a price that had been thought outrageous even for gasoline before the 1979 "crisis"), and he added, "We cannot allow D.O.E. to ignore the promises made to Congress in 1976."

F. Kevin Boland, a G.A.O. official, made it clear in his testimony, however, that this was precisely what was happening. D.O.E. was merely serving as "a conduit" for industry figures, he testified. "It never considers the equity of price or competitiveness in refineries." He added that G.A.O. had been stonewalled by the department. It had asked Secretary

Charles W. Duncan on December 4, 1979, to spell out D.O.E.'s position, but its request had been ignored. As a result, Senator Warner had made the G.A.O. report public without any D.O.E. rebuttal—in effect, a plea of no contest.

The second study to demolish Big Oil's free-enterprise myth and D.O.E.'s credibility was conducted by the staff of Representative Dingell, chairman of the House energy subcommittee. In a hearing on November 9, 1979, Dingell pointed out that heating oil prices had increased anywhere from 60 to 100 percent, depending on the time of comparison, from those charged in the previous year, "far outstripping the price of crude oil." He referred to the public's "confusion and anger" at having to pay such prices while major oil companies' third-quarter reports for 1979 were showing their tremendous 100- to 200-percent surge in profits.

Dingell stressed that industry witnesses and D.O.E. itself had always insisted in sworn testimony as well as in informal questioning that price increases resulted from increased costs. But this simply was not true. A memorandum prepared for him by the subcommittee staff showed that oil executives, when pressed, admitted that these "cost-only" assertions had all been lies.

In one public hearing, Texaco had admitted that its refineries' gross margin had increased from 9 cents to slightly more than 18 cents per gallon, a 102-percent increase. This was happening at a time when industry indexes showed that refiner costs had increased only 28 percent, "leaving 75 percent of their increase unexplained by cost increases," Dingell said.

D.O.E.'s phony public posture as the watchdog monitoring prices was exposed by Dingell's staff in this passage of its memorandum:

The staff interviewed D.O.E. experts on heating oil and asked how the companies priced their product. Although they admitted *they had never asked the companies,* they felt [the companies] performed elaborate maginal cost analyses—like the textbooks suggest. The D.O.E. staff explained that *the major refiners were responsible companies and would not gouge the consumer— therefore, they did not attempt to verify any of the numbers used in their various heating oil margin studies.* [Italics added.]

Having obtained this admission from D.O.E. that it never questioned anything but accepted whatever the oil companies gave it because they were all honorable gentlemen together, the Dingell staff set out to discover how heating oil prices were really determined. They closely questioned officials of some of the major oil companies with this result:

Contrary to their recent testimony before congressional committees, the major oil companies did no marginal cost analysis as the basis for their enormous price increases over the past year. Staff interviews by top officials of some of the major oil companies have revealed that most of the companies "charge what the market will bear."

In determining this maximum market tolerance, the companies cited as a justification a daily pricing service known as *Platts Oilgram,* published by McGraw-Hill. *Platts* reported high and low transactions for the previous day in major market areas. How did *Platts* know? "Apparently," the subcommittee memorandum said, "no one has ever investigated *Platts* to determine how they collect their price information." The investigators found, however, that *Platts*'s data might well be unreliable.

One executive told the staff that he was familiar with the *Platts* process—they merely call a few refineries and check the prices— they do not check with both the seller and the purchaser to determine whether a bona fide sale took place. He claims that some cargoes have been sold as many as four times [another version of the Daisy Chain], resulting in a jacked-up price that was accepted by *Platts.* He said that "of course it can be rigged."

Another "top official of a major company admitted to the staff that years ago he had rigged one of these pricing services."

The *Platts* quotes may help the industry in its effort to justify the unjustifiable, but the oil companies juggled prices to suit themselves even without this device, the subcommittee staff found.

The fear of the heating oil shortage [actually, there was no shortage, but the 1979 gasoline "crisis" had created the fear] has created a market that is particularly vulnerable to monopoly pricing, and this is what has occurred.

As an example, the staff cited interviews with Gulf Oil Corporation officials who admitted frankly that they did not base their prices on costs but on "what the market will bear." Between April 1978 and June 15, 1979, Gulf pricing decisions were made by district managers in Gulf's sixteen districts. After June 15, pricing authority was centralized in Houston. The manager of Gulf's Philadelphia district was especially proud of his performance in the freewheeling period when he had authority to set prices. He admitted that he had jacked up heating oil prices *twenty-eight times in nine months,* most of the raises coming between January and June of 1979. "The district manager informed the staff," the memorandum said, "that his objectives were to maximize revenues and to minimize costs in order to optimize profits." This district manager felt that, as long as he was not out of line with the rest of the market (here *Platts* data would help as justification), whatever price he set was a fair price. The subcommittee memorandum added:

It is interesting to note, however, that what the major companies have done in most markets is to leapfrog each other on price, essentially ratcheting the price up, with each company following the market that has no relationship to cost increases.

How much did this "ratcheting up" in price hurt the American homeowner who, unlike the driver, could not go from gas station to gas station seeking the best price? Imprisoned in his home, at the mercy of the system, he had to pay—and pay. The enormity of this most unconscionable of ripoffs was indicated in one thorough, detailed study made by the staff of Representative Rosenthal.

The study showed that in just thirteen months—from September 1978 through September 1979—American consumers had been overcharged $3.4 *billion* for No. 2 heating oil and middle distillates like diesel fuel. And this was only the tip of the proverbial iceberg, for the Rosenthal study ended just as heating oil and other middle-distillate fuels were about to take off in their most frenzied escalation since decontrol.

The Rosenthal staff's analysis was based on the industry's own figures. It relied for its statistics on D.O.E.'s *Monthly Energy Review* and *Sales of Fuel Oil and Kerosine in 1978*, and on the National Petroleum Council's report *Refinery Flexibility*. During its investigation, the committee staff kept in constant contact with sources inside D.O.E. both to verify the figures and to update them.

That research led Rosenthal's staff to this conclusion:

The refiners' margin for heating oil had increased from 7 cents to 21 cents/gallon by September 1979, and for diesel fuel had increased from 7.2 cents to 21 cents/gallon for the same period. Average operating and capital costs have remained relatively constant. Prevailing inflation rates were 11–12 percent during the period, the industry had undertaken only modest capital expansion, and labor costs had remained at less than the rate of inflation under the administration's wage/price control program. Because of relatively higher costs for fuels at refineries, we assume an overall cost increase of 20 percent from the NPC [National Petroleum Council] reported costs of 5.4 cents a gallon. This calculation yields an estimated cost average for 1979 of 6.5 cents/gallon and, consequently, a profit margin of 14.5 cents/gallon for heating oil and diesel fuel. The net refinery profit jump from 1.6 cents [on heating oil] and 1.8 cents [on diesel fuel] to 14.5 cents/gallon (allowing for operating cost increases) works out to a relative profit increase in excess of 800 percent for heating oil, and over 700 percent for diesel fuel.

Accordingly, it is estimated that the increased profit margins for these middle-distillate fuels yielded refiners *an additional $3.4 billion profit.* [Italics added.]

When Representative Rosenthal disclosed these findings at a hearing of his House Subcommittee on Commerce, Consumer and Monetary Affairs on January 20, 1980, the $3.4 billion ripoff in just thirteen months was a figure so stunning that it secured wide publicity across the nation. And the spontaneous outburst of public outrage indicated that New York's Mayor Koch had not been off the mark when he had told Stuart Eizenstat that he had seen nothing comparable to it since Vietnam.

Letters poured into Rosenthal's office from all sections of the country. They told of financial hardship, dangerous flirting with hypothermia, and burning resentment at the oil companies.

From Auburn, N.Y., a man wrote, "At night we set the Thermostate [sic] at 65 and during the day at 68 degrees, and we have to wear additional clothes to keep warm. I am 85 years old and it takes nearly all of my Social Security in order to pay for our fuel bill."

A woman in Weatherly, Pa., wrote that she and her husband were both retired and that 263 gallons of fuel oil had just cost her $234.06—$80 more for 40 gallons less than she had paid at the same time the previous year. "Why is it that fixed income people and middle-class workers always get it in the neck and their pocketbook?" she asked.

A retired eighty-second Airborne Division major wrote, "When one thinks of these older citizens that froze from lack of fuel . . . so that these corp.'s could garner huge profits, I have to question the administration's move. . . . I'm a Democrat. But Carter and his carpetbaggers must go."

A writer from Lamar, Colo., accused the administration of a "sellout" and denounced D.O.E. as "the worst Dept. that ever faced the tax paying public." And from Rockville, Va., a man wrote, "Anyone who says we have a free enterprise system is either stupid or getting a rip off from the profits. . . . I am frustrated and rather feel like the early Americans did just before the Boston Tea Party."

When I returned to New Jersey after covering some of these hearings in Washington, I encountered my fuel oil dealer friend and found that he was just as irate about Big Oil's dealings as Representative Rosenthal's correspondents had been. Perhaps he was even angrier than some because he saw from the inside what was happening and how badly some people were being hurt.

Angry myself, I happened to mention that I had just paid 97 cents a gallon for a delivery of fuel oil, and he stunned me with this bit of inside-the-industry scuttlebutt: "Do you know, Mr. Cook, what they are projecting? One—dollar—and—twenty—eight—cents a gallon! They figure the price will stabilize there because that's about all the market will bear."

We then talked about President Carter's recent action in drawing a line in the Persian Gulf and warning the Russians it could mean war if they crossed that line and tried to grab the Middle East oil fields. This really infuriated my oil dealer friend. "Do you think I'm going to have my son go over there and go to war for the oil?" he asked, his voice rising. "If we went to war for the oil and the oil belonged to the people, that would be one thing. But suppose we went to war and we won it and we got the oil, who do you think that oil is going to belong to? Exxon, that's who! Do you think I'm going to have my son go to war for Exxon and Company? If you do, you're crazy!"

Such was the mood of a badly hurt, suffering, infuriated —and helpless—people. The kind of mood that was to "bring down a presidency," as Representative Rosenthal had so perceptively foreseen.

My friendly oil dealer was accurate in his forecast about

heating oil prices. He had predicted almost to the penny, a year in advance of the event, the ceiling these prices would reach. Heating oil prices, which closed the 1979–80 season at 99.9 cents a gallon, went just past his $1.28 prediction to $1.30 in the winter of 1981. The world oil glut of 1981–82 brought them back to $1.23 at the start of the new heating season. And at this price the oil companies were charging far more for fuel oil than they were for more expensively processed gasoline, considering the 12-cent state and federal taxes gasoline had to bear.

In the late fall of 1981, I pulled into a gas station to buy regular gas at $1.219. As luck would have it, the fuel oil deliveryman who used to fill my tank before I switched to natural gas was putting heating oil into the underground tank of a building next door. I walked over and asked him, "Well, what's it costing today?"

"1.229," he said—a penny more for untaxed fuel oil than I was paying for 12-cent-taxed gasoline. Then the deliveryman looked at me and said, "Isn't it terrible?"*

---

* Tax situations vary from state to state, but in New Jersey, where heating oil is untaxed, these two prices meant that the oil companies, for beginners, were ripping off 13 cents a gallon for heating oil, which does not begin to require the refining gasoline does. The resulting windfall for Big Oil boggles the mind. According to the Fuel Merchants Association of New Jersey, between 1.2 billion and 1.4 billion gallons of No. 2 heating oil were sold in the state in 1981. Since the 1982 winter was far more brutal, consumption might be expected to rise. But even the minimum 1981 figure of 1.2 billion gallons, if multiplied by this off-the-top skim of 13 cents a gallon, yields an unjustified profit of $156 million from this source alone. It could well be more. By the end of February 1982, heating oil prices had dropped to $1.189 a gallon, but gasoline prices had fallen even more to $1.159 for regular gas. This meant that 15 cents a gallon, not 13, was being ripped off. A fuel oil association spokesman admitted, "The refineries have just shifted their profit center and loaded it on to fuel oil."

11

Wrong-headed

Policies

The Carter administration exhibited a positive genius for adopting wrongheaded policies that could only hurt the American people, enrich Big Oil with multi-billion-dollar windfalls—and in the end bring down the American economy in the worst recession since World War II. In the process, the administration ensured its own demise in the election of 1980.

The explanation for this misbegotten performance lies in the big-business roots, not the grass roots, of Jimmy Carter. His policies were founded, not on independent thinking, but on the kind of cerebration that is swayed by the dollar signs on the bottom line. As a result, the Carter administration's oil policy was based on two fundamental propositions: First, decontrol oil prices, letting the American market rise to the highest world levels as determined by OPEC and Big Oil; second, force the American people to conserve by this high-price scenario and, if that wasn't enough, impose a stiff surtax on imported oil that would drive American prices *above* the world going rate.

This policy of soaking the public until it hurt was seen as essential to reduce our imports of oil from OPEC and guarantee our national security so that we would not be caught in another crippling bind if the Arab states should cut off oil again as they had in 1973–74. (This thinking assumed as fact, of course, despite much contrary evidence, that the 1974 "crisis" was real, not a mere contrivance by OPEC and Big Oil to drive prices up and share the loot.)

It is not hindsight to say (I was writing this at the time) that Carter's policy flew in the face of common sense and that it ignored the most obvious needs—genuine competition in the energy field and the development of cheaper, not more expensive, energy sources to prevent an inevitable, ruinous inflation. High-priced petroleum products do not affect just the automobile driver and the homeowner using heating oil; they affect literally *everything* across the whole scope of the American industralized economy because *everything*, in one way or another, depends on a petroleum or natural gas base. The fertilizer for our farms depends on it; so do our farm machinery and the trucks that transport the farms' produce. Every steel plant, automobile factory, commercial establishment depends on it. So do our pharmaceutical houses, our plastics industry. The cellophane wrapper on a loaf of bread, the synthetics in our clothes, the asphalt on our streets—all depend on this energy base. Drive the price of that energy into the stratosphere, and you drive the price of everything else into the stratosphere along with it.

The Carter justification for a policy so purblind was that the United States, with its vast resources of coal as an alternative and its profligate use of energy as a habit, must drastically reduce its heavy imports of oil both to lessen its deficits in international trade and to relieve pressures on the world oil markets that adversely affected countries like France, Italy and Japan that had few natural resources of their own. But a low-priced scenario, by undercutting OPEC-posted prices, could have relieved these worrisome oil pressures without the concomitant domestic disaster. This could have been achieved by following two routes: establishing an honest and efficient D.O.E. to impose strict price controls on the oil industry, preventing outrages like the Daisy Chain and heating oil scams; and developing and using an abundant natural resource immediately at hand—

natural gas. It is a program that, carried to its logical extreme, might have ended up curbing Big Oil interests and treating them like public utilities. But why not? They are *the* basic public utility. We control utility prices, but we don't control the source on which such prices have to be based. It is idiotic to control one and ignore the other.

There is no evidence that the Carter administration ever gave even passing consideration to such a heretical, price-competitive alternative. All the evidence suggests, on the contrary, that it recoiled in alarm and horror at any proposal, any fact, that would discredit its doomsday crisis scenario.

The roots of the administration's thinking may be traced to two principal sources: the Trilateral Commission, of which Carter had been a member, and a series of energy studies commissioned by the Ford Foundation that had emphasized enforced conservation as the cornerstone of national energy policy.

In a 1974 report, the Trilateral Commission called for conservation efforts that would hold the increase in energy consumption below 2 percent for the next decade. It placed special emphasis on the necessity for the United States, by far the heaviest user of oil and the possessor of the most alternative resources, to curb its energy consumption to avoid draining world markets to the detriment of other nations' economies. To achieve this virtually static energy rate, the Trilateral Commission suggested, "in the case of industrial consumption the price mechanism should be the main incentive to saving energy." It added, "Limitation of civilian consumption will also be necessary. Here a balance will have to be struck between the price mechanism and the use of mandatory controls."

The Ford Foundation, in a 1974 report, *A Time to Choose: America's Energy Future,* developed in more explicit detail a similar program. It called this a "Zero Energy Growth Scenario" (ZEG). This called for zero energy growth after 1985. It suggested that stringent policies would be needed to raise the cost of energy and "limit its availability sufficiently for energy consumption to level off."

This study proposed that, to avoid imposing insupportable hardships on the poor, "energy stamps" similar to food stamps might be issued by the government—a suggestion that found an echo in Carter's 1979 allotment of $1.6 billion from the windfall profits tax to help those at or near the poverty level to pay some of their heating oil bills. (This was a woefully inadequate amount. It was not enough to cover everyone in need, and it was estimated at the time that even those who got help would not receive more than $200, barely enough to fill a small oil tank once during bitterly cold winters.)

In a second report issued in 1979, *Energy: The Next Twenty Years*, the Ford Foundation applauded some of Carter's policies. Like him, it favored price decontrol. "The primary energy-producing industries are workably competitive domestically," it wrote, "and the market—if it is allowed to operate—can be expected to produce, conserve and allocate energy with reasonable efficiency. . . . Price decontrol would, to some extent, stimulate domestic production of oil and gas."

The report also commented, "It is essential that investors in energy production know that in the long run, they will be paid at least as much for oil that they produce domestically as they and others will be paid for oil that they import." Again: "We suggest decontrol now, deferring for later consideration the possibility that *prices even above the world level should be imposed.*"* (Italics added.)

---

* This was exactly what President Carter attempted. In early 1980, he tried to impose by executive order a $4.62 fee on every barrel of imported oil, in essence raising it above the OPEC price by that amount. Rebuffed in the courts, he tried to get Congress to approve the surcharge. Interestingly, John C. Sawhill, himself a Trilateral Commission alumnus who had been brought back to Washington as deputy secretary of energy, wrote a ringing defense of Carter's import fee proposal on the Op-Ed page of *The New York Times*, May 26,

Though the Ford Foundation made many worthwhile suggestions for conservation through more energy-efficient devices, improved car mileage, more insulation, the improvement of the decrepit mass transit system, its calm assumption that there was competition in the energy field flouted the clear evidence of decades of experience. And the foundation's acceptance of the idea that domestic prices should be determined by world prices was in essence a ratification of the insatiably greedy deeds of OPEC. It was virtually a surrender of American independence, since foreign producers would be setting the benchmarks that would determine the impact of energy prices on the American economy.

One has to wonder what history those who drafted the Ford reports were reading, for the experience of more than fifty years shows a pattern of manipulation and control, the creation of repeated phony crises and price mark-ups after every one, the development of sudden, crippling "shortages" and their mysterious disappearance the instant the price objective was achieved. The record does not spell out *free enterprise*; it spells *cartel*. And the Standard Oil companies have always been at the heart of it.

The antitrust decision of 1911 broke up John D. Rockefeller's original Standard Oil trust, but the record shows that the huge Standard Oil companies, their directorates inter-

---

1980. Sawhill argued that the import fee would raise gasoline prices only 10 cents more a gallon, that it was necessary to reduce our imports of oil by 100,000 barrels a day and that it was essential to show good faith to our allies, many of whom were paying higher prices for gasoline than we were. Congress passed a resolution repealing the fee by overwhelming votes in both House and Senate. Carter, who had just won enough votes to guarantee him renomination for president, vetoed the bill, vowing he was acting on principle, even if he got only one vote to uphold. He didn't get much more. The House promptly overrode the veto by a vote of 335 to 34 amid an outbreak of hisses, groans, boos and mock cheers. It was the first time since 1952 that a Democratic Congress had overridden the veto of a Democratic president.

locking with the Rockefellers' Chase Manhattan Bank, continued to act in concert for their mutual interests. Mobil was originally Standard Oil of New York; Exxon, Standard Oil of New Jersey. Standard of California uses the name of Chevron and is also known as So Cal. Standard of Indiana markets under the name of Amoco. Rockefeller interests, through their own holdings augmented by those of the Chase Manhattan Bank in various arms of the old empire, still dominate much of the market.*

This is clearly shown in the lineup of the so-called Seven Sisters, the companies that compose the Arabian American Oil Company (Aramco), which has had virtually exclusive rights to Middle East oil. The Seven Sisters include Exxon, Mobil and So Cal, which means that three-sevenths of the combine is made up of Standard Oil–related companies. The others are Texaco, Gulf, Royal Dutch Shell and British Petroleum.

It is ironic—and illuminating—that the same scare tactics with which Americans became so familiar in the more recent "crises" of 1974 and 1979 were used as far back as 1920. In that year, the Standard Oil companies wanted to break into the Middle East and tap the rich oil fields then controlled by the British, and the propaganda tom-toms began to thunder. The American public was assailed with terrifying estimates that we were about to run out of oil. Gasoline supplies were so controlled and so restricted that some stations had to sell it in only one- or two-gallon lots. Congress and the State Department were whipped into a froth of indignation about this strangling of the American

---

* According to Ralph Nader's Public Citizen's Tax Reform Research Group, Chase Manhattan, one of the largest American banks and the most powerful banking link to Big Oil, not only paid *no income taxes* to the federal government in 1980, but had an effective tax rate of *minus 22.4 percent.* By contrast, an American worker earning $18,000 wouldn't have a tax credit coming to him but would be paying his income tax at a rate of 24 percent. (See column by Ruth Simon, Op-Ed page, *The New York Times*, February 28, 1982.)

economy; and the British, influenced by this uproar, let the American companies get their first foothold in the Middle East. Just incidentally, the price of gasoline doubled to 35 cents a gallon, the equivalent of $1 in today's currency.

The pattern was repeated with minor variations through the years. In the mid-1950s, Venezuela offered for lease its rich Lake Maracaibo oil fields. Vigorous small American companies got into the action, and some of them struck it extremely rich. Huge and highly productive fields were developed by such little-known firms as San Jacinto and Hancock Oil. This new flood of Venezuelan oil posed a threat to the majors' control of supply and price, and the cartel went to work on the Eisenhower administration, waving the flag and calling for the protection of the American oil industry from this foreign invader. Oddly enough—strange how it always seems to happen—Big Oil had a friend close to the throne, Treasury Secretary Robert B. Anderson, a Texas oil man on whose advice Eisenhower relied. The result: The Eisenhower administration imposed a set of stringent controls. Foreign imports were held down to 12 percent of domestic production, and only companies with refineries—mostly the majors, of course—were allowed to import. This action killed off the independents who had developed some of the richest Maracaibo fields, and they were swallowed whole by some of the larger American companies having the necessary refineries. This action, which drained our own petroleum resources instead of using readily available foreign supplies, made little sense, but it did ensure the majors' control of the supply and price structure.

Then came the 1970s—and new threats. Libyan independents brought in new gushers and drove down the price of crude oil to $1.30 a barrel. Market analysts predicted the price might fall to $1, and Big Oil board rooms shuddered. Naturally, it never happened. Abroad, the Libyan government was pressured to curtail production; in the United States, independents that had been giving the majors competition thanks to cheaper imported oil suddenly found their supplies cut off or reduced to the point where they could not stay in business. By the time the 1973–74 "crisis" arose, there were a lot of closed and abandoned gas stations around the country.

When Egypt and Syria invaded Israel in the Yom Kippur War on October 6, 1973, OPEC suddenly rose like a phoenix from the ashes. OPEC had been organized as far back as 1960, but it had been a paper tiger without clout or influence. Yet two days after war broke out, officials of the major oil companies met with representatives of OPEC in Geneva, and OPEC came striding out of that meeting wearing seven-league boots—a miraculously born power that overnight dominated the world of energy.

The OPEC sheiks announced that no oil would be shipped to the United States or the Netherlands, Israel's best friends, until the Israelis surrendered all the Arab lands they had seized in the 1967 war. At the same time, OPEC imposed a schedule of new, uniform prices that boosted the cost of oil dramatically. There was no more nasty talk about the possibility of dollar-a-barrel oil. Saudi Light, which had been selling for $1.80 in late 1970, jumped to $5.11 and then to $11.65 on January 1, 1974—a sixfold increase in four years, with more than $8 of the boost coming in three months between October 1973 and January 1974.

There is no evidence that Big Oil was outraged. In fact, there is much evidence to the contrary. Big Oil apparently knew months in advance of the event that the Arabs were going to war with Israel. Frank Junkers, president of Aramco, visited Saudi Arabia's King Faisal on May 3, 1973, and he received such a strong tip from Faisal and his advisors that he felt certain war was coming. Junkers later said that he passed this warning on to Exxon in New York and Standard in California. What happened next is intriguing. For the next six months, from May to October, Saudi Arabian production increased 18 percent over what it had been in the first quarter, and the major oil companies built up their inventories at what soon turned out to be bargain basement prices. By the end of 1973, inventories of the major companies had risen 5.5 percent over the figure at the end of 1972.

Thus the Arab embargo should have had relatively little effect on them, especially as important volumes of oil continued to flow from non-Arab sources like Iran and Nigeria. As was to be the case later, in 1979, there were widespread reports that deliveries reaching Atlantic ports were greater than the industry reported, that domestic production was being held back and wells capped, that tankers had been slowed to a crawl at sea and that empty tanks in abandoned gas stations were being filled to the brim to store excess gasoline.

In the midst of this apparent plenty, there was suddenly a "crisis"—a gasoline shortage so severe that motorists for the first time had to wait in blocks-long lines at service stations. Limitations were put on the amount of gas each driver could buy. Most stations stayed open for only a few hours a day, a couple in the morning and a couple in late afternoon. Flags went up: a red flag for no gas, a yellow representing a limited number of gallons, a green signifying plenty. In this "crisis," there were no green flags; all was chaos. And much evidence says it was all unnecessary.

Signs abounded that indicated there was no real shortage of gas. In my area of New Jersey, several residents who lived near abandoned gas stations began writing letters to the newspapers describing how tanker trucks would roll up during the night and fill the empty gas tanks. Representative James J. Howard, the congressman from our district, made a tour of some of these abandoned stations, measuring stick in hand, and the *Asbury Park Press* carried a picture of him as he measured a tank and found it full of secretly stored gasoline.

There was another aspect of this "crisis" that seemed so

unnatural as to be weird. The New York metropolitan area, a media nerve center for the nation, seemed to be extraordinarily hard hit. In the early winter of 1974, a young couple working on a television project drove down from Boston to see me. When they arrived, they were amazed at the severity of the gasoline shortage in our area. "What is happening around here?" the young man asked. He said that he had had no trouble getting gasoline until he hit the New York area; then he found service stations closed for the night, heard about the long lines and the limited, rationed sales. He couldn't believe what was happening.

Another case in point: A friend of mine and his wife had been in the habit for years of taking their house trailer to Florida to spend a winter month. With this hideous new gas "crisis" upon us, they had qualms about going. Finally, the husband, a good mechanic, welded an extra gas tank to his trailer, rigging it so he could switch to this reserve if his main tank began to run dry. Since he worked at one of the army bases in the area, he managed to get both tanks filled with gas, and he and his wife started out. Later, after they had returned from Florida, he was still in a kind of amazed shock as he described for me what he had found on the road.

"As soon as we got out of New Jersey, there was no problem—no problem at all," he said. "We were getting down to the Maryland border when I saw a gas station open. My main tank was getting low, and I thought, 'Well, I better pull in and see if I can get a few gallons.' So I pulled in, and the gas station attendant came up and said, 'You wanna fill 'er up?'

"I was flabbergasted. I said, 'Well, sure, if you can. Don't you have any limits on gas around here?' 'Naw,' he said, 'we got plenty of gas.' And you know, we didn't have any trouble all the way to Florida—you could get all the gas you wanted anywhere. No trouble except in New Jersey. I couldn't get over it."

What amazed him, and me, is that there we were sitting on top of some of the largest refineries on the East Coast—Exxon's huge plant at Bayway, for example—and yet it was *this* very area, New York and its metropolitan suburbs, that seemed to have been crippled the worst, almost as if by design.

Such was the story of the 1974 OPEC embargo "crisis." Prices rose 100 percent or more as the oil companies sold those huge inventories at the new OPEC benchmark, cashing in many times over on what they had paid for the oil. It is no wonder, as Commissioner Jacobson had written to Senator Bradley, that 1974 was the most profitable year Big Oil had ever had until 1979 came along.

The myth of the Iranian shortfall and the engineered phony gasoline crisis gave Big Oil a bonanza that eclipsed anything on which it had feasted before. The "crisis" faded almost as quickly as it had risen, but the bonanza kept right on going. *The Wall Street Journal* was puzzled in early November by a situation that seemed to defy the rules of a free-enterprise society. The *Journal* reported that "global oil production is plentiful," that American consumption had declined and that there seemed to be an excess of production over consumption amounting to one million barrels a day. Supply and demand in such a case should have brought prices down; instead, they continued to zoom as if there were still a critical shortage. OPEC, impervious to the laws of free enterprise, had increased prices by 10 to 12 percent in October; it was planning another increase in December. These actions and the psychology of fear—fear of another "Iranian shortfall," another "shortage," more OPEC price hikes—were what continued to drive prices up in an over-supplied market, the *Journal* concluded. It was not a very convincing explanation.

The balance sheets of Big Oil were more revealing. Wall Street analysts conceded that every time OPEC raised its prices, Big Oil profited. It profited because it used the OPEC benchmark to set higher prices on the crude oil it produced elsewhere in the world; it profited because every time OPEC raised its prices, Big Oil immediately jumped the price of the billions of gallons it held in inventory. When Saudi Arabia boosted its price of $2 a barrel to $26 in late January, 1980, and Iran raised its figure to $31, Big Oil profited hugely from all the frantic buying of the previous fall that had so puzzled *The Wall Street Journal.*

When the oil industry reported its quarterly profits, the truth about what had happened in what the *Journal* called the "erratic" oil market of 1979 began to emerge. Exxon's net profit for this year of travail and "crisis" topped $4 billion, and its sales reached a record $84.35 billion—more than the gross national product of Sweden. And this was only the beginning. The first-quarter reports for 1980 disclosed how hugely Big Oil had benefitted from the late 1979 escalation of prices that had defied the law of supply and demand.

Exxon's first-quarter profit increased 102.1 percent and came to *$1.92 billion,* the largest one-quarter profit ever made by a corporation in American history. The rest of the Big Oil companies showed similar gains. Mobil's net for the quarter more than tripled to *$1.38 billion.* Texaco's doubled to slightly more than *$1 billion.* Amoco's first-quarter gain was 64.6 percent, or $576 million net. Smaller companies showed booming first-quarter profits ranging from 33 percent to 103 and 109 percent above those of the previous year. It was an unholy outburst of greed and profit that dwarfed earlier increases that critics had called "obscene."

While the oil companies were reveling in these doubly obscene profits, the rest of the economy was decimated. Inflation was in double-digit figures; interest rates were soaring. By spring 1980, automobile manufacturers had laid off 212,500 workers—and this was only the beginning. They were soon to lay off thousands more, and the ripple effects spread to other industries supplying the big automakers. The housing industry, thanks to zooming construction costs and interest rates, was in a shambles. Analysts were already predicting the worst recession since World War II with unemployment mounting to almost 9 percent and more than 9 million workers unemployed.

Such were the disastrous effects of a misguided policy that poured the wealth of the nation into the pockets of one industry at the expense of all others. Had Jimmy Carter been the populist he posed as being when he ran for president, he might have read the history of Big Oil without blinkers on and adopted more rational policies. But from the first days of his presidency he had embraced a doomsday script tailored to the interests of Big Oil. To do so, he had to ignore—as the Ford Foundation, the Trilateral Commission and the double-domes influenced by Big Oil had ignored—the simple fact that we had within our borders an energy resource truly awesome in its possibilities: natural gas. If the truth about this had ever been allowed to escape from the dungeon in which it was confined, Carter's entire "We're perched on the precipice" script would have lost its credibility; and since this credibility was the administration's first priority, anyone who questioned it risked his official neck.

The first to feel the axe was Dr. Christian Knudsen, who headed an energy research team of the Energy and Research Development Administration (ERDA) within D.O.E. He had worked for years for Exxon in synthetic fuels research before joining the Federal Energy Administration, D.O.E.'s predecessor, in May 1974. After the Carter administration took office, Dr. Knudsen was placed in charge of the Market-oriented Program Planning Study (MOPPS). Its task was to determine the supplies and cost of the nation's oil, gas, coal and electricity.

Until he received this assignment, Dr. Knudsen had been highly respected and honored for his work in ERDA. On January 1, 1977, he had been given an achievement award and a prize of five hundred dollars. The certificate that went with the cash praised him for "consistently outstanding performance and noteworthy achievements." Less than four months later, Dr. Knudsen had become a pariah within the department.

His trouble began when his research team, organized February 1, began to compile the basic statistics that President Carter would need for his forthcoming energy speech. When Dr. Knudsen and his staff, using U.S. Geological Survey (USGS) reports and industry-supplied data, looked at natural gas resources, reserves and cost, they quickly discovered that natural gas was the most abundant and cheapest energy source that we had.

This discovery conflicted with the administration's, and Big Oil's "crisis" scenario. Secretary Schlesinger and his deputy, O'Leary, had been quoted in the press since February as saying that we were running out of natural gas and that this could be disregarded as an energy resource. Thus Dr. Knudsen's conclusion that we had enough natural gas to last the nation well into the next century touched off a series of internal shocks in the higher echelons of D.O.E. The full story was exposed months later in hearings conducted by the Senate Energy and Natural Resources Committee, chaired by Senator Metzenbaum, with Senator John Durkin of New Hampshire playing an active role.

The picture of Dr. Knudsen that emerges from the hearings is that of an unsuspecting scientist who had walked into a political bear trap. In late March, as details about his study began to become known in the department, anxiety manifested itself. One of his superiors suggested he call in an outside consultant, Dr. William Vogely a Penn State University professor, to double-check his findings. Vogely had been a highly respected research economist in the Department of the Interior before he left government service for his university post. Dr. Knudsen contacted him, and Vogely spent April 5 studying Knudsen's graphs showing supply and cost curves. He also examined the statistics on which the curves were based. Dr. Knudsen later testified that Vogely "thought we had done a very good job on the curve, except that he thought we were not showing as much gas as was there." Vogely also thought that any price above $1.42 per thousand cubic feet (Knudsen's figures were higher) was too much. "So he felt he would prefer a lower cost and a larger resource curve."

This Ploy Number One having failed to discredit Knudsen, Ploy Number Two had to be tried. Dr. Hugh D. Guthrie, director of the Division of Oil, Gas and Shale Technology in ERDA, told Dr. Knudsen he might bring in an expert from Shell Oil to examine the troublesome Knudsen supply curves. Dr. Guthrie, it so happened, had been employed by Shell for thirty-three years before landing a top spot in ERDA. In the hearings, Senator Durkin found it significant that Guthrie had dismissed Vogely's conclusions "with the sweep of a hand" and had decided to bring in someone from Shell "to substantiate . . . his [own] previous statements."

More curves were thrown around now than exist in a baseball pitcher's repertoire. The object was to support the Carter-Schlesinger-O'Leary thesis that the nation had only ten years' supply of natural gas and that it would quickly became so scarce and costly that its price would zoom to between $5 and $6 per thousand cubic feet. It was suggested that Dr Knudsen try using a curve along these lines prepared by the Stanford Research Institute (SRI). This SRI curve was apparently based on information supplied by Gulf Oil, but there was no indication what methodology had been used or on what statistics the curve was based. Knudsen asked his superiors in writing for this background information, essential to determine how valid the SRI projections were. His request was ignored.

When this series of events was outlined during the hearings, it irked Senator Durkin.

Now, Doctor, it would seem that the SRI model is a black box; no one knows what's in it, no one knows the assumptions, no one knows whether they're valid except Gulf Oil, and a few people that have it.

"For a novice like me, it would seem like a shell game outside a racetrack going on: The only problem is you don't know whether there is a pea under any one of the shells."

Senator Durkin pointed out that, if Dr. Knudsen's figures were correct, we had enough natural gas to last for forty years instead of ten, and this meant "there is enough natural gas in this country to break the back of OPEC and bring the price of oil down, isn't there?"

This was the whole point, and Dr. Knudsen acknowledged that this was precisely what was at stake. He saw the possibility that, if natural gas were allowed to compete with oil, "oil may have to seek a [price] parity with gas"—an idea that is still anathema to Big Oil, which is doing its damndest to drive natural gas prices up to a parity with oil.

Dr. Knudsen had presented his final report to a convocation of his ERDA superiors on April 7. It was a long session; there was detailed discussion because everyone knew these figures would be important in view of the upcoming Carter speech. Yet, there was no serious effort to tear the Knudsen conclusions apart, and Dr. Knudsen left the meeting with the false impression that his analysis had been fairly well received. He was soon to be disabused.

Deputy Secretary O'Leary was later described as having "hit the roof" because the natural gas figures, once and twice revised, weren't coming out the way they *had* to come out to suit the administration's game plan. Finally, on April 15, 1977, Dr. Knudsen was called into a conference room for what Senator Durkin later described as the "wirebrushed" meeting. In more common parlance, he was being taken to the woodshed.

Describing the experience, Dr. Knudsen later testified that "it was an emotional meeting, a heated meeting." Dr.

Guthrie, he said, told him that his natural gas curves "were extremely misleading" and that "they would be extremely sensitive in the public eye, subject to misinterpretation, and might have a serious effect on energy policy." One of the top executives at this "wirebrushed" session said that he didn't think Dr. Knudsen knew anything about natural gas. When Dr. Knudsen protested that Dr. Vogely had checked his research and found it too conservative, Dr. Guthrie, ignoring the fact that ERDA itself had recommended Vogely, interjected angrily that "Mr. Vogely was totally discredited, his estimates were discredited by the industry." He didn't say what industry, but he didn't have to. It was obvious Dr. Vogely had been knifed by Big Oil interests.

After this acrimonious session, Dr. Knudsen was doomed. The man who had been honored for his meritorious service in January was fired from his job as director of MOPPS in April. (There were indications that the decision to get rid of him had been made as early as April 12—three days before the meeting.) With Knudsen and his heretical supply graphs out of the way, the doomsday figures the administration wanted were forwarded to the White House. On Monday night, April 18, President Carter went on national television to make what Senator Durkin described as his Chicken Little "the skies are falling" speech to scare the wits out of the American people.

12

The Natural Gas

Gusher

The drilling rig stood tall behind an old tenant's shack on the flat, steamy plain of Pointe Coupee Parish near Baton Rouge, La. Suddenly, there was a deep rumbling in the bowels of the earth, and a huge gusher erupted from the bore hole, shrouding the drilling rig in a fog-thick cloud of steam—and natural gas.

The date was August 13, 1977, one that may go down in energy history with August 29, 1859, when the Spindletop well in Titusville, Pa., ushered in the age of petroleum.

Chevron USA (the production arm of Standard Oil of California) had been drilling its Walter C. Parlange Well No. 1 in Pointe Coupee Parish. It had sunk casings to a depth of 20,000 feet. Then it bored deeper. When its bit hit a level of 21,346 feet, it apparently penetrated a geopressured zone holding an incredible quantity of natural gas. The instantaneous eruption, much like the blowing of a rich oil well, was dramatic proof of what some of the best geologists had been contending—that the United States had such huge supplies of natural gas that there never should have been an energy crisis.

Chevron's strike demonstrated the fabulous richness of the natural gas deposits that underlie the long, vast stretch of the Louisiana-Texas coastline and that extend in a 120-mile-long, 40-mile-wide strip inland across Louisiana and into the Texas Panhandle. This inland section, centered around Baton Rouge, is known as the Tuscaloosa Trend; the

138

coastal waters are the Miocene Trend. Both contain innu-
merable reservoirs of natural gas.

It had cost Chevron $5 million to drill its Walter C.
Parlange well, but when the bubble burst, natural gas came
pouring out at the astonishing rate of 142 million cubic feet
a day. In just twenty-nine days, Chevron sold to Florida Gas
some 3.6 billion cubic feet—enough to supply thirty-one
thousand homes for a year—at an estimated return of
$5,469,916.

This pay-back of drilling costs in just twenty-nine days
was a phenomenal record, and it shook the oil industry.
Chevron, after recovering its costs, capped the well, and no
one knows whether, in those twenty-nine days, Chevron
made other sales besides those to Florida Gas. This and many
other details are cloaked in the legal dodge of proprietary information, about which Chevron has remained tight lipped.

There was even much dispute about exactly what it was
that Chevron had hit. Chevron tried to play down its find;
Dr. Paul H. Jones, a Baton Rouge hydrologist, played it up.
Dr. Jones, a veteran of thirty years' service with the U.S.
Geological Survey, is generally acknowledged to be the fore-
most authority on the natural gas resources of the Tuscaloosa
Trend. He had long argued that geopressured aquifers lying
thousands of feet deep in this onshore trend across Louisiana
and into Texas contained as much as 50,000 trillion cubic
feet (TCF) of gas. Since national consumption is about 20
TCF annually, this meant that, if Dr. Jones was right, the
nation had enough natural gas in the Tuscaloosa Trend
alone to last for centuries.

When the Parlange well blew, Dr. Jones went to the
scene as soon as he heard the news. It was well that he did,
for Chevron contended that it hadn't hit a geopressured zone
at all, that it had just struck an abnormally rich pocket of
methane, the principal component of natural gas; that what
had blown above the top of its drilling rig wasn't steam,
but gas and that it had flared off the gas immediately to get

the well under control. A Chevron engineer at the scene, however, had taken a series of color photographs, and he gave these to Dr. Jones. They told a different story.

The first photograph in the series clearly showed a cloud of steam (identified as such in the engineer's handwriting on the back) enveloping the top of the rig. The fourth picture in the sequence contained a notation on the back indicating that Chevron hadn't flared off pure gas immediately as it had said. The notation read, "Steam and gas discharge over the pit immediately after the successful ignition of flare. Late August 1977."

Why would Chevron want to deny it had made a discovery comparable to Sutter's gold? Some skeptics suggested that, since natural gas prices were controlled at the time, Chevron wanted to hold back production until prices rose. Chevron said it had sold its gas for $1.45 per thousand cubic feet. Chevron also said it had had to cap the well because the casing had been damaged by the blowout and was no longer usable. But there may well have been another, more compelling reason.

According to Dr. Jones, Chevron didn't have a license to drill into geopressured or geothermal zones; in addition, when it brought in its monster, it began to drain gas from peripheral areas. This led to a series of lawsuits. One of those who sued was Walter C. Parlange himself. Parlange, who lived in a nearby 1750s plantation house, owned the land on which Chevron had drilled, and he sued the company for $20 million, contending that the gas Chevron had sold belonged to him because it came from geothermal sources not covered by the leases he had signed with Chevron. Other nearby landowners also sued because they weren't being paid royalties for gas that, they argued, was being drained from the deep reservoir under their lands. The suits were all settled out of court for a rumored $15 million.

In the meantime, Chevron had moved its rig a few hundred yards away from its Walter C. Parlange Well No. 1.

There it drilled a second well to a depth of 20,000 feet. It described this well only as "highly productive." Chevron argued that this second, 20,000-foot-deep well showed it was tapping only conventional supplies, but Dr. Jones asserted the well was simply tapping a gas cap created by the blowout.

Whatever the facts hidden behind the screen of proprietary information, the spectacular blowout of the Parlange No. 1 well shook up D.O.E., Congressional committees in Washington, and a lot of experts who had persistently ignored the natural gas potential. For example, the Ford Foundation's 1974 report only mentioned natural gas in passing; its 1979 report just made dubious reference to it. These reports had concentrated on methods of increasing oil production, on the development of our large coal reserves and on nuclear power. Such was the conventional wisdom of the time.

In this heavily charged big business and political atmosphere, anyone who tried to call attention to our natural gas bounty risked his neck. One of the most distinguished scientists in the field took that risk. Dr. Vincent Ellis McKelvey had had a long and distinguished career in government service. He had received the Distinguished Service Award from the Interior Department and the Rockefeller Public Service Award, among many honors. He had been chief geologist of the U.S. Geological Survey, and in 1971 he had been named its director. It was a post he still held in 1977 when Mount McKelvey in Antarctica was named after him.

Then, in this same year of 1977, Dr. McKelvey made a speech in Boston. He said that natural gas reserves were so huge they amounted to "about ten times the energy value of all [previous] oil, gas and coal reserves in the United States combined." Dr. McKelvey, relying on recent seismic studies of the geological survey, added that the geopressured zones in Louisiana and Texas alone held some 24,000 trillion cubic feet of gas—the equivalent of about 4 trillion barrels of oil, roughly twice the conventional estimate of ultimate world petroleum resources. All that was needed, he added, was the development of the technology to tap these reserves.

Dr. Knudsen had made himself persona non grata with his far more modest forty-year estimate of natural gas supplies—and had been demoted. Dr. McKelvey's offense was far worse; and so, in the aftermath of his Boston speech, the man for whom a mountain in Antarctica had just been named was ousted as director of the geological survey.

Now retired in Florida, Dr. McKelvey insists that he doesn't know to this day whether he was fired as the result of his Boston speech. The speech, he says, had been cleared in advance with the Secretary of the Interior, and all he was told afterwards—this after his long years of distinguished service with the geological survey—was that "it was felt it would be better for the department and best for the geological survey if I retired."

Such were the hazards of bucking the Carter–Big Oil "crisis" scenario designed to keep prices up at the expense of the American people. As Dr. Jones later commented to me, "The whole subject [of natural gas] is looked upon as a threat to the price structure. The synfuels program would go down the drain if they produced this."

To help understand all that is involved, some technical explanations are in order. Organic sediment deposited millions of years ago as ocean tides ebbed and flowed over this vast area became buried miles deep under the forming land mass. Geopressured aquifers were created as the whole weight of land mass above bore down upon water-bearing porous rock formations, exerting pressures ranging from 6,000 to 15,000 pounds per square inch depending on the depth. Temperatures, too, rose from 230 to 400 degrees Fahrenheit the closer the aquifers lay to the earth's molten core. This combination acted like "a pressure cooker," in Dr. Jones's words, literally boiling layer after layer of sediment into petroleum and methane gas. The methane dissolved in the steaming hot brine solution, trapped in its aquifers below layers of impervious rock, like shale. Some free-floating gas escapes through faults in the upper rock formations and forms into domes within 2,000 to 3,000 feet of the surface, but the much larger, richer bounty still lies trapped in the deep aquifers. Once a drill punctures the rock shield above such an aquifer, the tremendous pressures built up over centuries of time send an explosive burst of steam and gas spouting to the surface—just what Dr. Jones contends happened at the Parlange well.

For decades, drillers had been tapping the higher domes where gas and petroleum were often found mixed. To wild-catters seeking oil, the gas was just a nuisance, and for years it was flared off, hellish lights glaring above the oil fields as trillions of cubic feet of gas were wasted. Such idiotic expenditure of a precious energy resource has now been stopped, but until the 1970s most wildcatters had little use for gas. Sometimes geopressured gas blowouts would wreck their casings, and they would simply cap the well and leave it. At other times, seeking oil and finding gas, they would cap the well in disgust and prospect somewhere else for the more valuable petroleum. Many of these old, abandoned well sites in Texas are now being taken over by a new breed of wildcatters who are drilling deeper, with heavier casings and equipment, seeking the buried gold that for so long was scorned.

My attention was first called to the significance of the natural gas story while I was on vacation in 1980. I met an old friend who had done some work for one of the largest drilling companies in the West. "Why don't you do something about natural gas?" he asked me. "We've got enough of that to last for a thousand years." Startled, I asked what he meant. He cited the experience of the corporation with which he had been connected. "They've drilled a couple of holes to twenty thousand feet looking for oil," he said. "Instead, they hit such huge pockets of natural gas that the pressure blew their rigs right out of the ground."

This started me on my research. I found that a number of articles had been published, mainly dealing with the Parlange blowout, but the major media had made almost no attempt to follow the story or try to assess its potential significance in the national energy picture. Bryan Hodgson had written a thorough article, strictly limited to scientific aspects, for the *National Geographic* in November 1978. *Barron's,* the *Christian Science Monitor* and *The Wall Street Journal* had published articles on the Parlange spectacular. But none of these had contributed much to a public understanding that natural gas supplies could be so huge as to warrant the entire rethinking of the national energy problem.

One respected researcher who had spent a year tracking down the natural gas story told me, "This is the major uncovered story in the world. All sorts of new fields are being opened up, with twenty to thirty million cubic feet a day coming on stream." Yet official figures, he said, did not reflect this increment; they remained static.

Dr. Jones said the same thing. "The D.O.E. people are dead set against natural gas," he told me when I talked to him in late May 1980. "Yet the difference between producing the natural gas that we have in volume or not producing it is the difference between the United States' going down the drain for $100 billion a year, which is what we'll be paying OPEC."

Official figures, Dr. Jones added, showed "an increment of only 1.7 trillion cubic feet in Louisiana last year [1979], but the figures I get show an increment of fifty trillion cubic feet in the Tuscaloosa Trend alone." He thought that his original prediction that we had reserves totaling 50,000 trillion cubic feet (enough to last for 2,500 years at the current annual 20 TCF rate of consumption) was now "too conservative."

Other experts agreed that the resource base was enormous, far greater than official figures showed. Geologist Oscar Strongin had told the House Subcommittee on Energy and Power in 1977, "All of the investigations of the geopressured zones of the Gulf Coast virtually unanimously agree that it potentially contains resources that can contribute materially to our natural gas supplies." He had added that, in some areas, sediments were 50,000 feet thick. Yet, he said, the petroleum industry viewed these deeper, massive geopressured zones "primarily as curiosities while it has gone about its business of finding and developing conventional deposits of natural gas."

Tests were showing, however, that these geopressured "curiosities" might contain far more gas than anyone had ever predicted. One such experiment was run by D.O.E. itself in Tigre Lagoon, deep in the humid coastal marsh of southern Louisiana. The well was called Edna Delcambre No. 1, and Bryan Hodgson, who was on the scene, described the results in his *National Geographic* article.

It is producing as much as 10,000 barrels of water a day from a sandstone aquifer 13,600 feet below, where pressure is almost 11,000 pounds per square inch and the temperature is 240°F. The gas is collected in an elaborate manifold system, and the water is forced by its own pressure down another well bore for disposal 2,500 feet underground.

Major oil companies have made similar tests in utmost secrecy. But Edna Delcambre No. 1 is in the public domain. And the results are dramatic. Its production is 150 percent higher than originally predicted.

Dr. Denton Wieland, the geologist in charge of the experiment, told Hodgson he had expected to get 20 cubic feet of gas from each forty-two-gallon barrel of water, but the well was yielding, instead, 50 cubic feet per barrel.

D.O.E., having made this dramatic discovery, capped the well and made no attempt to draw broader conclusions from the data it had collected. But its significance was not lost on others.

Dr. William M. Brown, of the Hudson Institute think-tank, cited the Edna Delcambre results in a 1979 study as an indication that previous estimates of the amounts of recoverable gas in the Louisiana-Texas Gulf region might be far too low. The Edna Delcambre production, he wrote, indicated that "in addition to the hoped-for dissolved methane some free gas might also be extracted. This unanticipated result created a bit of excitement as it gave support to the optimistic speculation that *most* of the geopressured reservoirs might contain some free gas in addition to the dissolved methane. Moreover, it is believed that the producible amount of free gas might be greater than that of the dissolved gas."

After analyzing all possibilities, Dr. Brown concluded conservatively, "Even without the deep gas there are respectable estimates that producible U.S. resources of conventional natural gas also can exceed 1,000 TCF [a fifty years' supply]." The Edna Delcambre test and other drilling along the Gulf Coast suggested to Dr. Brown, however, that not only were the brines "fully saturated" at drilling depths of around 22,000 feet, but lower brines like those that Dr. Strongin had talked about might be not just saturated with methane but "super-saturated" since pressure and heat increase with greater depth.

This careful and conservative Hudson Institute analysis suggested, therefore, that conventional resources in the Gulf Coast region alone could supply the nation with enough natural gas to last for at least fifty years and that huge, untapped geopressured zones might have the potential to extend this supply for centuries with advancing technology.

Dr. Jones believes that the technology conventionally used is inadequate to recover the amounts of gas available. "The trouble with conventional thinkers," he says, "is that they don't understand hydrology." He explains that, in the past, standard procedures for drilling a gas cap required that pressure below the cap be maintained so water would not invade the well. When so much gas had been removed that the pressure dropped, water flooded in and wells were abandoned. Dr. Jones believes that only a fraction of the gas in such wells was ever retrieved. He is convinced that water should be pumped from deep aquifers in enormous quantities as swiftly as possible. This rapid reduction in pressure, he reasons, would result in an upward rush of gas bubbles mixed with brine. In some instances, the gas would free itself from the water completely and refill the gas cap or form a new cap. Dr. Jones has patented this process and, as this is being written, is hard at work perfecting it and getting financial backing.

Even without such advanced technology, there was a veritable explosion of natural gas discoveries in 1979–80. Announcements of new finds and the development of whole new fields came sometimes at a pace of three or four a day. And, in many cases, natural gas was flowing at mind-boggling rates.

In August 1980, Superior Oil revealed it had discovered a "significant" new field in the Sabine Pass Block 3 area in the Gulf. It was significant, indeed. A wildcatter considers himself fortunte if he hits a well that flows at a rate of 4 million cubic feet a day. Superior's Sabine Pass field was producing 27 million cubic feet a day from six wells—and five other wells that had encountered gas-bearing sands were being developed.

In December, Mobil announced it was developing a field in Mobile Bay, twenty-seven miles south of downtown Mobile. Mobil estimated the field contained between 200 billion and 600 billion cubic feet of gas. About the same time, Exxon reported it had found a field in the Gulf ninety-five miles south of Intracoastal City, La., with multiple gas sands ranging from 8,200 to 9,700 feet, an extremely thick and rich reservoir.

Amoco weighed in with the announcement that it had

tested three more wells in its Port Hudson field in the Tuscaloosa Trend, twelve miles north of Baton Rouge. The field, it said, was the largest yet discovered in the Tuscaloosa. The three new wells were flowing at a rate of 22.5 million cubic feet a day, and Amoco said six other wells in this field were producing 55 million cubic feet a day—a total of 77.5 million cubic feet.

Dr. Jones emphasized the significance of this development. The Port Hudson field, he said, is a veritable monster "with 700 feet of sands full of gas." He estimated that the Port Hudson field is worth a billion dollars a square mile and that it will be producing for more than twenty years. He illustrated by comparing it to the Hollywood field in the Tuscaloosa Trend. Hollywood, he said, has nowhere near the depth of gas-bearing sands that Port Hudson has. Yet "that field was first opened in 1957, and it has produced a trillion cubic feet of gas in an area six miles long and two and a half miles wide. It is still producing, and pressure in the wells is still so high that it indicates there's still a lot more gas there."

Official estimates of what natural gas wells will ultimately yield have always been from 15 to 20 percent too low, Dr. Jones said. (Others told me the same thing.) There always seem to be more pressure and more gas in the wells than were accounted for in the original estimates. Taking this into consideration, the nation, it would seem, has plentiful supplies of natural gas for centuries to come. Prudhoe Bay on the Alaska North Slope has an estimated 26 trillion cubic feet of gas waiting to flow south once the Alaskan gas pipeline is completed—and no one knows how much other natural gas other Alaskan reservoirs may hold. Thus huge discoveries along the Gulf Coast and the Tuscaloosa Trend that have been described here are a mere sampling of the burst of new discoveries in the region. And this area, phenomenally productive though it is, is just one of three blockbuster zones known to exist in the lower forty-eight states. The other bountiful areas are the Anadarko Basin of Oklahoma and the Overthrust Belt along the Rockies.

The broad Anadarko Basin that stretches across Oklahoma and down into the Texas Panhandle has been one of the richest and most productive natural gas areas in the nation. By 1980, 3 trillion cubic feet of gas had been found at levels below 15,000 feet, and only 2 percent of the deep-lying sediments had been tested with the driller's bit. Shallow wells were producing additional heavy volumes of gas, and Robert A. Hefner III, the pioneer deep-driller of the Anadarko, was convinced that the basin held more natural gas than conventional estimates had allotted to the entire United States.

Hefner, who heads the GHK Company in Oklahoma City, has been drilling in the Anadarko since the early 1970s. In his early drilling, he brought in thirty successful wells, and his company participated with other concerns in developing others. He is regarded, therefore, as the principal expert on the resources of the Anadarko, just as Dr. Jones is on those of the Tuscaloosa Trend.

The Natural Gas Policy Act of 1978, which lifted controls on gas drilling below 15,000 feet, spurred a frenzy of activity in the Anadarko. Prior to the passage of the act, most drilling had been done by wildcatters like Bob Hefner, but beginning about November 1979, the majors began to pour their multibillion-dollar resources into explorations in the area.

By early 1980, the deep wells already drilled and those targeted for drilling speckled a map of the Anadarko like a chicken pox rash. Shallower wells were so numerous that the *Oil & Gas Journal,* the bible for the industry, didn't even attempt to spot them on a map it published, testament in itself to the abundance of the Anadarko's resources.

Indeed, natural gas exists in such quantities in the basin that normal wildcatting hazards have been all but eliminated. The accepted rule of thumb in the industry used to be that if a wildcatter struck it rich in one well out of ten, he was lucky. But these odds don't hold for the Anadarko. Bill Dutcher, an associate of Bob Hefner, explained when I talked to him in 1980.

Overall, during the early 1970s, about 60 percent of all the wells we drilled below 15,000 feet were producers. In the last two years, we've drilled twenty deep wells, bringing in fifteen producers. We haven't hit any real barn-burners yet, but we have developed some really strong producers. These wells will produce anywhere from one million to five million cubic feet a day, and they'll produce for the next twenty or thirty years—at a gradually decreasing rate, of course.

In the Anadarko, Dutcher explained, drillers have what he called "a bail-out effect." He cited GHK's experience in drilling what was at the time the deepest well in the world in Washita County, Okla. The well was drilled to a depth of 31,441 feet in 1974. It found no gas at that level, but as the drilling equipment was brought back up the hole, it struck a productive zone at 13,000 feet.

Before 1979, independents had done virtually all of the drilling in the Anadarko, but then the big oil companies— Exxon, Conoco, Getty—began to pour in money and equipment. The result was a boom-town atmosphere reminiscent of California in the gold rush days of the last century. Housing was at a premium around Elk City; there was no unemployment. "Since November 1979, it's been all systems go," Bill Dutcher said enthusiastically. "The independents and the big companies will still be drilling the Anadarko twenty to thirty years from now. We're really excited at the way things are going."

He is not the only one excited. Once-sleepy villages located in the Overthrust Belt that runs along the Rocky Mountains from Colorado to Wyoming have become boom towns. Discovery after discovery in recent years has convinced many veterans of the drilling game that the Overthrust Belt contains tremendous resources of oil and gas—

but especially natural gas. Like the Tuscaloosa Trend and the Anadarko, it has been only partially explored.

The Overthrust Belt was formed eons ago when the Rockies heaved themselves up above the western landscape. In the process, layers of rock thrust outward, overlying and crushing lower levels where oil- and gas-rich sediments formed. For centuries, this potentially rich lode was overlooked, but with the development of new seismic techniques for studying lower strata in the earth, interest perked up.

In 1975, the American Quasar Oil Company, of Fort Worth, Tex., hit big pay dirt in the Pineview field near Coalview, Utah. Geologists say that a wildcatter considers himself the world's luckiest man if he hits oil-bearing strata 100 feet thick. Given this perspective, one can readily appreciate the liquid gold that Quasar found—a lode 450 feet thick with oil and natural gas. Even richer was the Painter Reservoir, discovered by Chevron in southern Wyoming in 1977. It had strata 300 feet thick loaded with oil, topped by almost 700 feet of natural gas.

These discoveries were just a prelude to the natural gas eruption of 1980 and 1981. In June 1980, Amoco struck it rich in the Overthrust Belt in northern Utah and southern Wyoming. Three productive zones in just one field were flowing at a rate of 40 million cubic feet a day—enough to supply 123,000 American homes with all the energy needed for heating, cooking and hot water. And this was only the beginning. A month later, Amoco brought in two more wells along the flanges of the Overthrust. One was fifteen miles north of Evanston, Wyo., the other, four miles southwest of Evanston. The first well flowed at a rate of 10.6 million cubic feet a day, the second at 10 million—enough to take care of another 60,000 American homes.

Amoco's luck continued into 1981. In May, it announced one of its richest discoveries. Its No. 1 Champlin 458-F well, the third development well in its field near the East Anschutz Ranch field in northeastern Utah, delivered 3,300 barrels of an oil-like condensate and 18 million cubic feet of sweet gas a day from depths between 12,800 feet and 13,350 feet. The zone was not only exceptionally thick, more than 450 feet, but the sweet gas was an additional bonanza. Much of the gas coming from the Overthrust contains sulphur, an impurity that has to be processed out of it; but sweet gas is so pure it can go right into the pipeline.

Huge strikes continued. Texas International Company's stock jumped on the New York Stock Exchange with its report in February that it had encountered exceptionally rich and thick oil and gas sands in its Eloi Bay field in St. Bernard Parish, La. One analyst described Eloi Bay as one of the three most important discoveries in the United States, ranking it with the Williston field in Montana and the Overthrust Belt in Wyoming. The early reports of the importance of the Eloi Bay field were confirmed in November 1981, when Texas International disclosed that a Houston engineering firm, after an independent study, had concluded that the field had a potential of 2 trillion cubic feet of gas. It was believed capable of producing 200 million cubic feet a day when fully developed, and arrangements were being made for a $25 million pipeline to transport the gas to the Louisiana intrastate market.

Texaco brought in two wildcat wells in the Gulf off the Louisiana coast. One was described as flowing at a rate of 12.6 million cubic feet of natural gas a day. Apache Corporation hit the richest gas well in its history in Wheeler County, Tex., a part of the Anadarko. The well flowed at a rate of 24.5 million cubic feet a day. Washington Gas Light hit two "barn-burners," to use Bill Dutcher's term, in the Anadarko. One developed so much pressure that it collapsed the well casing and had to be abandoned. It had had an uncontrolled flow of 60 million cubic feet a day before the blowout. A second well drilled nearby to tap the same reservoir hit pay dirt at a depth of 16,568 feet, was kept under control, and was delivering 27 million cubic feet a day.

Such huge additions to the nation's natural gas supply were not confined, however, to the already partially explored and known productive regions like the Tuscaloosa, the Anadarko and the Overthrust Belt. Natural gas was being found in sections of the country where no one had ever thought to look for it—in the Northwest Mist region of Oregon, a state that had never produced a whiff of natural gas before 1980; in Michigan, where wells were thundering in with reported flows between 18 million and 28 million cubic feet a day; in North Dakota, Ohio, West Virginia, Tennessee, Pennsylvania and New York. As my researcher friend had told me when I first started to investigate the hidden natural gas story, "Anyone who has looked for natural gas has found it."

Of special and growing interest were indications that an eastern overthrust belt exists along the Appalachians just as it does along the Rockies. This potential has been hardly tested, but initial drilling has given rise to optimism that here is another vast and potentially invaluable resource.

West Virginia and Tennessee, for example, are literally honeycombed with shallow wells. In November 1980, Presidio Oil of Denver and American International of New York put up $18 million to buy mineral rights around Charleston, W.Va. The amount of money and the limitations on the purchase were striking indications of the potential value of the area. The two firms spent their millions to buy from a Houston trust just 90 percent of the mineral rights going down to a depth of only 3,500 feet. The trust still retained the deeper—and potentially far more valuable—rights. Why would two companies put up $18 million just for 90 percent of shallow rights? A Presidio spokesman explained.

The purchase includes 109 working wells and rights to explore 130,000 surrounding acres. This whole area is pincushioned with shallow wells. We find most of the gas at depths of only 1,500 to 2,500 feet. This shallow gas is so extensive in this area that we have a record of 90-percent success in drilling. This is not just our own experience, but the experience of other drillers in the area. These shallow wells are slow producers, but the total volume of gas coming from them is mighty impressive or we wouldn't have put up all those bucks.

The reason the Houston trust wanted to retain all those deep-drilling rights in West Virginia may be found in the experience of the Columbia Gas System. In July 1979, Columbia Gas struck one of the richest wells ever tested. It was in Mineral County, W.Va., and it flowed at the fantastic rate of 88 million cubic feet a day. The discovery spurred a lot of drilling by major oil companies, but they failed to produce any wells of comparable size, though some quantities of natural gas were found.

From West Virginia, this Appalachian Overthrust Belt extends into Tennessee. There, in May 1981 the Towner Petroleum Company, of Lorain, Ohio, signed a twenty-year agreement to supply natural gas to Tenneco and East Tennessee Gas Company, a subsidiary of Tenneco. At the time, Towner's Beauregard field in Overton and Furness counties had eleven wells producing about 750,000 cubic feet of natural gas a day. But Towner also had thirty-six capped wells in Tennessee's Cumberland Plateau. The wells contained gas Towner would have liked to sell but couldn't because, it said, "Tenneco is the only possible buyer," and Tenneco didn't need the gas.

This account of the natural gas discoveries of 1980 and 1981 could go on *ad infinitum*. In pursuing the story, I found out, as my researcher friend had told me at the beginning, that new fields producing millions of cubic feet were being brought on stream almost daily. Yet, as he had also warned me, official figures did not reflect this enormous increase. Officials continued to act as if the gas wasn't there, confirming by their obtuseness Dr. Jones's statement that D.O.E. was dead-set against natural gas.

How dead set became apparent on January 7, 1982, when D.O.E. released what it called accurate, corrected figures for 1980. This was the year when so much drilling had taken place that the industry ran short of rigs and piping, and drillers had to wait sometimes for months to get needed equipment. This was the year, along with 1981, when mammoth new wells were being discovered and developed. And this was the year, according to D.O.E., in which the nation's natural gas reserves actually *declined.*

Proved reserves, D.O.E. said, fell from 201 trillion cubic feet in 1979 to 199 trillion cubic feet in 1980. They fell despite the fact that the nation had produced and consumed only 18.7 TCF in the year, a figure below the 20 TCF it often used.

The official report didn't surprise me; it merely shocked me by its brazenness. Dr. Jones had forecast this result when I talked to him in December 1980. He had told me then, "I've had to pull out of contact with D.O.E. altogether because they take your figures and misrepresent them. I don't like to use the term, but I don't know what you can call it but a conspiracy." At the same time, a harried spokesman for the Louisiana Department of Natural Resources, when asked if wells were being capped in the Gulf of Mexico to hold down supplies and wait for decontrol and higher prices, waffled all over the landscape but finally conceded in a voice that sounded scared, "You almost think there is a conspiracy somewhere."

In view of what had happened, a big-power conspiracy to falsify the report on the nation's natural gas supplies makes sense. Dr. Knudsen and Dr. McKelvey had felt in very personal ways the muscle of this hidden power. Big Oil had been conspiring throughout its history to create periods of shortages, to induce panic with energy "crises"—all with the purpose of boosting prices and profits astronomically. And now Big Oil was faced with a crisis itself: Natural gas was selling at less than half the price of No. 2 fuel oil. If this continued, if the nation realized it had abundant supplies of natural gas, there would be serious *competition*, a dread development in Big Oil's version of free enterprise.

As a result, an all-out advertising and publicity campaign began in 1980 to steer homeowners away from natural gas— and keep them buying heating oil at Big Oil's prices. The Heating Oil Council, in an advertisement in *The Wall Street Journal*, proclaimed, 98% OF AMERICAN HOMEOWNERS WOULD BE BETTER OFF STAYING WITH OIL HEAT. In my own Asbury Park area, a fuel oil dealer took out an advertisement headed, NATURAL GAS. HERE TODAY . . . GONE TOMORROW? And in New York a fuel oil distributor passed out a flier reproduced from an advertisement in *Newsday*. In bold, block type it warned, NATURAL GAS PRICES EXPECTED TO TRIPLE.

It all seemed a part of an orchestrated Big Oil campaign to discredit the natural gas alternative; to force rapid decontrol that would bring prices up to a parity with No. 2 fuel oil—and thus eliminate dangerous competition. Literally billions of dollars are at stake. Billions in the continuing ripoff of the American public in heating oil and gasoline prices. Billions more in Big Oil's dominance of synfuel projects to liquify or gasify coal—something that can be done at a profit only if energy prices are kept sky-high. And they can't be kept that high if natural gas—the most abundant, cleanest and cheapest energy resource we have— is allowed to ruin the scenario.

Big Oil has to keep that genie capped in a bottle; and so, as this is written, it is playing all over again its favorite game of scarcity to justify price decontrol, on the phony premise that only decontrol can spur drilling and give us more gas. Its real object is to bring those natural gas prices up to a parity with No. 2 heating oil. Which means the end of any meaningful competition. Which means the ratification of Big Oil's ironclad monopoly over every facet of American energy.

This is what is at stake in the raging battle of the 1980s over natural gas decontrol and pricing. It is a battle that is going to affect in multiple and fundamental ways the nation's future.

Recipe for

13

Catastrophe

On November 9, 1981, several thousand protestors marched through the streets of Chicago and picketed in front of the Conrad Hilton Hotel where Big Oil moguls of the American Petroleum Institute were holding their annual convention. The pickets carried signs reading STOP BIG OIL, STOP NATURAL GAS DECONTROL, FREEZE PRICES, NOT SENIOR CITIZENS! and FARMERS SAY: DON'T FREEZE US OUT!

The mass protest was an effort to stop the ultimate Big Oil ripoff—the decontrol of natural gas prices with the purpose of making this most abundant and cheapest of fuels as expensive as No. 2 heating oil.

The demonstration had been organized by the Citizen/Labor Energy Coalition (C.L.E.C.), headed by William W. Winpisinger, president of the International Association of Machinists and Aerospace Workers. Protestors had come from thirty states by bus, train, car and on foot. Young and old, black and white, factory workers and farmers and housewives, they had rallied to protest gas price decontrol that, it was estimated, would at least double—perhaps triple—heating costs for 55 percent of American households (some 43 million residential users) that depend on natural gas. It would also have a devastating impact on commercial establishments, major factories like those in the automobile industry, and even the farmers in Texas who depend on

158

natural gas, not just for fertilizer feedstocks, but for the pumps that irrigate their acres.

Such was the issue as the delegates to the A.P.I. convention looked down from their Hilton Hotel rooms on the marching, sign-waving protestors. Some carried signs reading 30% MORE? NO WAY! HONK. And HONK AGAINST HIGHER BILLS! Cab drivers passing the Hilton read the signs and some of them honked loudly in agreement—a miniature reminder of the cacophony of honking horns that had mocked the Nixon White House on the night of the Saturday Night Massacre when President Nixon had tried to halt a probe into his activities by firing investigator Archibald Cox and Attorney General Elliot Richardson.

On Sunday, November 8, the day before the mass demonstration in front of the Hilton, protestors had held a rally at which Representative Dingell, chairman of the House Energy Committee, had assured them, "They'll pass decontrol over my dead body!" At this meeting, C.L.E.C. released a report claiming that political action committees of A.P.I. members had spent at least $3.1 million in federal campaign contributions since 1979 to further natural gas decontrol. The study charged that fourteen major oil companies alone would reap an additional $80 billion in revenues—and $40 billion in additional profits—once natural gas prices were allowed to blow sky high like the Parlange well.

Such were the high stakes for Big Oil on the one hand, for the nation and its people on the other. The whole issue had been exacerbated by the hard-right, big-business ideology of the Reagan administration.

President Reagan had taken office vowing to deregulate both oil and natural gas. His rationale for such action was the industry's rationale. Lifting all controls, Big Oil said, would spur drilling for more oil and gas; it would let free enterprise work its will and the market place set fair prices. It was a script that conveniently ignored basic facts: that we had a cartel-controlled, not a free-enterprise market, and that, even with controls in place, drilling had been going on at a maximum pace throughout 1980. Decontrol couldn't possibly spur more activity. It would just permit a cartel-controlled industry to set higher prices.*

This was precisely what had happened when President Reagan, in one of his first acts in office, lifted all the remaining controls on petroleum products. The White House expected prices for gasoline to rise no more than 3 to 5 cents a gallon over a period of several months. Big Oil shattered that myth in just ten days. Some companies boosted gasoline prices 15 cents a gallon even though there was such an oil glut at the time that some refineries had had to cut back or shut down. So much for the law of supply and demand.

Heating oil prices had tracked right along behind gasoline prices, and by March 1, 1981, in our New Jersey area, heating oil was selling at $1.279 a gallon—the same price as regular gasoline at some stations. "Isn't it awful?" the secretary of one fuel oil dealer said to me when I asked her about prices.

Reagan lucked out on this one. Oil prices gradually retreated. The deepening recession that closed plants and put millions of workers on unemployment lines cut fuel consumption and reduced driving. Higher-mileage cars required far less gas, and there was a worldwide glut of such proportions it could no longer be completely controlled. Yet the example of what had happened with heating oil decontrol and the increasingly hard times had raised to a boil the

---

* The oil glut of 1981–82 led some Reaganites to say, in effect, "See what happens when you remove controls." Yet Reagan's decontrol hadn't spurred extra drilling as it was supposed to do. *The New York Times* reported on March 21, 1982, "The Hughes Tool Company's weekly tabulation of rig activity hasn't shown a gain this year. The number of active rigs is down 619, or 13.6 percent, since January 1, and American refinery utilization stands barely above 60 percent."

opposition to tampering with natural gas controls. Reagan hesitated.

The hesitation was caused not by ideological but by practical political concerns. Reagan was embroiled during the first part of 1981 in a battle with Congress to get his first big-defense, big tax-cut budget approved, and his own Republican leaders in Congress did not want so divisive an issue as natural gas decontrol injected into the free-for-all. Once postponed, natural gas decontrol became an even more ticklish issue. The 1982 midterm Congressional elections loomed ahead, and even Reagan's diehard legislative followers shuddered at the prospect of running for re-election if they should have to face an electorate incensed by doubled heating costs.

Thus, there was only one way around the impasse, one way to avoid the kind of public outrage that had been demonstrated in Chicago—to accomplish by devious and subtle means what could not be done openly. Instead of going to Congress for approval of decontrol, Reagan could accomplish much the same purpose by executive fiat—by "backdoor" decontrol, handled through D.O.E., that would lift ceilings on various, specified categories of natural gas.

This involved extensive tampering with the 1978 Natural Gas Policy Act that President Carter had rammed through Congress by twisting every arm that was twistable. Rep. Toby Moffett, the principal spokesman for congressional delegations from the Northeast and Midwest, sections of the nation that would suffer most from higher gas prices, had led one of the bitterest fights in recent congressional history on the floor of the House—and had failed to block passage of the act by a single vote. The memory of that fierce contest made many legislative leaders shudder and wish they could avoid the whole divisive issue.

What the 1978 act did was this: It lifted all ceilings on natural gas produced from new wells drilled below 15,000 feet, it set up extremely complicated price-fixing formulas for various types of gas produced at higher levels and it provided for the gradual decontrol of gas prices, letting them rise at the inflation rate until 1985 when all ceilings would be lifted with one important exception. Gas from "old" wells drilled before 1977 would be still controlled—and this represented about 40 percent of the nation's annual production. Complete, immediate deregulation—the aim of Big Oil— would lift all restraints on the wellhead price of gas flowing from fields like the Hollywood field in Louisiana that had been producing since 1957 and had paid for itself many times over. Production from such wells under immediate decontrol would be allowed to claim the highest market price, and there would be no cheap-gas cushion to soften the sudden, crushing impact on the American consumer and American industry.

Could such a policy be justified? Did the industry need it to enable it to produce more natural gas? Or was this uncontrolled, high-priced scenario just one more example of the insatiable greed of Big Oil, regardless of what happens to the country?

Anyone who examines the record would have to conclude that the industry does not need the extravagant profits it is seeking. During the hearings on Dr. Knudsen's "wirebrushing," Senator Metzenbaum, who apparently had some private information, asked one D.O.E. witness whether he had ever been told that a major oil company "had opened a new gas field in Texas, sold gas for 9 cents per thousand cubic feet, on which they made money, but they were not too anxious to continue exploiting the field?" The witness couldn't recall having been given this information, though Metzenbaum understood that he had.

While 9 cents per MCF (thousand cubic feet) is doubtless less than dirt cheap, other evidence says that natural gas, unlike petroleum, costs relatively little to produce and distribute. The congressional investigator mentioned at the start of this book—the man who had visited a huge drilling platform in the Gulf in 1977 and saw all those capped wells—told me that he subsequently got into a friendly discussion over a few drinks with the boss of the drilling crew. The drilling boss admitted that, if the major oil company for which he worked wanted to produce all the gas it could, it could do so at a price of 18 cents per MCF— and still make a handsome profit.

Still too low a price? Perhaps. A higher appraisal, but one which won't please Big Oil much more, comes from David Schwartz, a Washington consultant who was an expert with the Federal Power Commission (FPC) from 1965 until 1978 when President Carter rolled it into the Federal Energy Regulatory Commission (FERC), under the wing of D.O.E. Schwartz cites the last gas study made by the FPC's staff in November 1978. The staff analysis showed that natural gas would be fairly priced at one dollar per MCF. "This was taking into consideration all costs, leasing, drilling, every-thing—and then allowing a 17-percent margin for profit," Schwartz explained.

The FPC, composed of Carter's industry-oriented appointees, decided to raise the figure to $1.52 per MCF. "They arrived at this figure by making allowance for taxes the companies said they paid," Schwartz explained. "No previous commission had ever granted such an allowance, and the staff had no way of knowing what taxes the com-panies paid, if any, because the companies refused to show us their books. Even the $1.52 price wasn't enough for Carter. When he came in with his Natural Gas Policy Act, the price was rolled up to $2.07"—in other words more than double what the Power Commission's expert staff had concluded would be a fair return, allowing for all possible expenses and a 17-percent profit margin.

When one adds to this the fact that Chevron admitted it sold the natural gas from its Parlange "barn-burner" for $1.45 per MCF—and still paid off its leasing and drilling costs in just twenty-nine days—one begins to get some idea of the kind of enormous profits even Carter's $2.07 guaranteed natural gas drillers.

It is difficult to argue that drillers did not have sufficient price incentives under the provisions of the 1978 act. The record shows that, not only has drilling been fast and furious, but the profits have been staggering. Dr. Jones points out that the last FPC study showed there were eight thousand producing gas reservoirs in the Gulf Coast region. (Texas officials told a congressional committee that the state had ten thousand working gas wells in 1981). The returns are enough to make one's head spin.

"These wells are paying for themselves in three or four months," Dr. Jones said. "After that, the principal expense to the drillers [during the twenty or more years the wells will continue to produce] is the payment of a 25-percent royalty fee to the landowners on whose property the gas has been discovered. I know personally of three cases in which landowners—each owned a large tract, of course—are getting $5 million a month in royalties."*

Yet a still higher return is needed? Let's not be ridiculous.

The best proof that no additional price incentives are necessary to spur gas exploration is to be found in Bureau of the Census data. This shows that between 1974 and 1978, despite the far stricter controls in force during the period, producers invested more money in gas exploration and development than they did in oil.

"This was so despite the fact that natural gas regulated prices were far lower than oil prices," Edwin Rothschild, director of the Energy Action Educational Foundation in Washington, wrote in *The New York Times*. "Why? Mainly

* Bob Hefner, the pioneer Anadarko deep-driller, is the man who was principally responsible for getting all ceilings lifted on gas produced below 15,000 feet. Hefner, who now receives $9 per MFC for his deep production, candidly admits that a deep well costing $12 million to drill can generate $500 million in revenues. (See *The New York Times*, November 29, 1981.)

because *the actual cost of producing gas is one-third the cost of producing oil*—all of which means that profits in every dollar invested have been greater for gas than for oil." (Italics added.)

The prices and profits have been leaping from year to year at a rate of about 20 percent under the 1978 gas act. The lowest-priced "old" gas has been selling for about $2.09 per MCF, but the price creep allowed for different categories of gas has resulted in an average price of about $2.50 per MCF. Prices have been kept this low in many instances because long-term contracts with producers, some of them signed in the late 1960s, often contained no escalator clauses and set hard, continuing prices of less than $2 per MCF. New, decontrolled gas from below 15,000 feet, however, is being priced at $9 per MCF, and the only thing that has kept overall natural gas prices from zooming out of sight has been the ability to "roll in" this extremely expensive gas into the overall stream with "old" gas that remains under controls.*

Yet the Reagan Administration, oblivious to public protests and congressional opposition, remains determined to remove all such safety valves and let the public suffer. It is a two-faced action that abandons decontrol publicly while trying to insure that it goes forward by devious means.

---

* Long-term contracts signed by some major producers indicate the price that natural gas could be produced for at a profit. Phillips Petroleum complained that it was selling some gas from its Texas fields for 19.7 cents per MCF. Phillips was locked into long-term contracts with two utilities, and a Phillips spokesman complained to *The Wall Street Journal* that the utilities' attitude was, "We've got you stuck, and we're going to keep you hung." Texaco, the second largest natural gas producer, complained that it received an average of only 98 cents per MCF for its production in 1980. Granted that the wild inflation of recent years has changed dollar values drastically, it seems nevertheless reasonable to assume that astute corporate managers would not have locked huge companies like Phillips and Texaco into such long-term deals unless they felt confident of making a profit at such low prices. The jump from 19.7 cents or 98 cents to $9.00 for much of today's unregulated production can hardly be justified by inflation and changed dollar values.

On March 1, 1982, the White House announced, "After extensive consultation with congressional leaders and groups representing producers and consumers, the President concluded that much needed changes to the Natural Gas Policy Act of 1978 would overload an already heavy legislative agenda."

Put in plain English, this meant that the administration didn't have the votes in Congress to do openly what it wanted to do. *The New York Times* pointed out that what little legislative support there had been for deregulation had "evaporated as the economy continued in recession and the elections moved closer." The *Times* also speculated that another factor had been the abnormally cold weather in January. This had raised heating bills for homeowners to new records. In Washington, D.C., the typical monthly residential heating bill had climbed 50 percent in January to $172.

The White House's aboveboard announcement coincided with an end run around Congress through the regulatory process. Charles M. Butler, a Texas attorney whom Reagan had made chairman of F.E.R.C., started promulgating new rules—and planning others—that would take the lid off the prices of large quantities of gas.

Butler's first maneuver in mid-December 1981 had doubled the wellhead price of gas produced from reservoirs lying under 300 feet or more of water. He and his commission set a new price of $5.10 per MCF, compared to the previous price of about $2.50. Then Butler proposed raising the ceiling price of a new category of medium-deep gas— that produced from wells between 10,000 feet and 15,000 feet in depth. Gas produced from this range from onshore wells would be allowed to increase to $3.85 per MCF or possibly to a much higher figure if Butler and his commission decided to boost the price 200 percent as they had done in some previous rulings. These two actions, energy analysts estimated, would affect pricing of at least one-quarter of all gas sold in the nation.

Even these two moves didn't satisfy the industry-oriented Butler. He let it be known that he wanted to loosen controls on even larger volumes of the so-called "old" gas that makes up 40 percent of the nation's gas line mix. Furthermore, he wanted to increase the rates that gas pipelines would be allowed to charge for transportating the fuel. The full impact of these last two proposals wasn't clear because the F.E.R.C. chairman hadn't indicated the size of the price hikes he sought, but all four speeded-up deregulation actions could be guaranteed to accomplish just two things if they went into effect: give Big Oil a tremendous new windfall running into billions upon billions of dollars, and ream the American consumer both in direct energy costs and in the inflationary pressures resulting from higher prices of every product with a natural gas base.

What would such costs be? It was typical of the insensitivity of the Reagan administration to the plight of the average American that Butler himself confessed he had no idea—and, apparently, didn't care. When reporters tried to find out the volumes of gas that would be affected by his deregulation moves and their potential cost, the F.E.R.C. chairman said, "We simply don't have any estimate of those things at this point."

This admission provoked some sharp criticism. Edwin Rothschild denounced F.E.R.C. for "rushing headlong to decontrol gas prices without collecting adequate information. It is one of the most disastrous plans ever conceived against consumers." And Rep. Philip R. Sharp, the Indiana Democrat who heads the House energy subcommittee that would have to consider any decontrol legislation, chided Butler for his "inadequate knowledge" of the potential effects of his policies.

The threat of Butler's sweeping "back door" decontrol led

House leaders to introduce a resolution denouncing his action and expressing "the sense of the House of Representatives that the Federal Energy Regulatory Commission should take no action to accelerate the decontrol of wellhead natural gas prices." Dingell, who presented the resolution, was joined by Dan Rostenkowski of Illinois, the commerce committee chairman, plus Sharp, Moffett, Rep. Richard L. Ottinger and Rep. Don Fugua of Florida, chairman of the House Science and Technology Subcommittee. Later, some one hundred members of the House became cosponsors of the resolution.

In a press conference February 24, 1982, announcing the introduction of the resolution, Dingell declared that Butler "must be stopped." He referred to the bitterly cold 1982 winter and said.

In fact, as things now stand, in my home state of Michigan, the average annual gas bill increased 41 percent over the last two years from $497 to $704. It is not at all uncommon, in this cold winter, to hear of bills well over $1,000 in a single month! And this administration wants those prices to go still higher. Our analyses indicate that accelerated decontrol of natural gas would double consumers' bills. Thus . . . this year's nationwide average residential gas bill of about $770 per year would rise to more than $1,500!

Though it is impossible to put an accurate dollar figure on what the Reagan administration's various "back door" deregulation moves would cost if they were all put into effect, there can be no question that full decontrol would be a catastrophe for the American economy. Rothschild's Energy Action Educational Foundation in a long, analytical report in February 1981 put the full cost of decontrol at $626 billion and called it "A Price Americans Can't Afford." The study said

Residential consumers will pay hundreds of dollars more for natural gas every year than they would if controls were continued. Across the country, industries will be crippled—and many small businesses will be bankrupted—by the sharp and unavoidable price increases. Decontrol will cause an increase in unemployment, a decline in productivity, and a big boost to the inflation rate. Decontrol will accelerate and intensify the drain of billions of dollars each year from the gas-consuming regions to the gas-producing regions. It will bring enormous hardship to natural gas users, and unprecedented gains to the major oil companies and the producing states.

Analyses like this bring screams from Big Oil spokesmen who try to portray Rothschild and Energy Action as wild-eyed alarmists, but both the director and his organization have been increasingly accepted in less partisan circles as responsible critics. Even the *Oil & Gas Journal,* the outstanding industry publication, ran a favorable profile of Rothschild and commented, "If he weren't so good at what he does, no one would care."

It is pertinent and instructive, therefore, to consider in some detail the impact of full decontrol as Rothschild and Energy Action see it. Between 1981 and 1985, their report found, the nation would pay that extra $626 billion under full decontrol split this way: "Residential users would pay an additional $165 billion; industrial users would pay $237 billion more; commercial users would pay $85 billion more; and electric utilities would pay $119 billion more than under continued controls."

Decontrol would add another three to five points to the inflation rate because, the report said, "for every dollar increase Americans pay in their heating bills, they will pay three dollars more for price increases in goods and services that consume natural gas. Industrial use accounts for 38 percent of total natural gas usage; commercial use accounts for 14 percent; and utilities using natural gas to generate electricity account for 19 percent."

In 1985, the average Northern worker heating and using gas for other purposes would have to spend four weeks wages just to pay his gas bill (not including transportation and distribution charges) compared to one week's wages if controls were still in place in 1985.

Even under a schedule of accelerated decontrol—and Rothschild at the time the report was prepared was not contemplating anything as radical as what F.E.R.C. is attempting—an additional $370 billion would be imposed on all classes of consumers between 1981 and 1985.

Events since this report was prepared have only tended to confirm some of its major findings. *The Wall Street Journal* in February 1982 described the devastating industrial impact of the rapid rise in natural gas prices. Spokesmen for U.S. Steel, the nation's largest steelmaker and its largest industrial user of natural gas, were worried because they anticipated that its current natural gas bill of between $400 million and $500 million would double in the next three years. Such an increase would almost equal the profits of $504.5 million that U.S. Steel had earned in 1981. The huge steelmaker had cut the amount of energy used to produce a ton of steel by 13 percent, but it was still using 100 billion cubic feet a year and it had paid 20 percent more for natural gas in 1981 than it had in 1980. It planned to substitute a much dirtier fuel, coal, for natural gas wherever possible. Nevertheless, Walter Jackson, U.S. Steel's director of energy, saw rising energy costs as handicapping the steel industry's efforts to meet foreign competition. "As the price of energy goes up, it will put an even greater squeeze on industry in this country," he said.

The squeeze will be worse for some industries than for steel, because for them there is no substitute for natural gas. The Owens-Illinois Incorporated glass bottle factory in North Bergen, N.J., is just one example. The plant had had to close one of its four furnaces and lay off 120 workers as a result of the recession, and the plant's manager, John Elliott, remarked gloomily that higher gas prices were "the very last thing we need right now." He explained that natural gas is the only suitable fuel for the plant's manufacturing process

—the fuel that allows the most precise control of temperature in glass making, the only fuel clean enough to come into contact with materials used for beverage containers. Despite the most strenuous conservations efforts, he said, the company's natural gas bill had jumped to $125 million in 1981 from $26 million in 1972. In March 1982, energy costs became so high that Owens-Illinois was forced to close down the entire North Bergen plant, firing 500 workers.

"Industrial users are going to get whacked hard," said L. William Sessions, vice president of energy and environmental quality for American Can Company. The firm's containers are used for food and beverages, and, as in the case of Owens-Illinois, natural gas is the only fuel clean enough to be used in their manufacture. Natural gas already accounts for some 20 to 25 percent of the operating costs at some American Can plants, Sessions said.

Textile businesses, like glass and can manufacturers, have to use natural gas. There is no substitute. Jack Elam, a vice president of Cone Mills Corporation, explained that natural gas is used in two processes. Denims and corduroys are singed with a flame to burn off extra-long cotton fibers, and a flame is used to dry the special resins and chemicals applied to the fabric. "Coal and oil leave residues in the fabric and change color," Elam said. Since 1972, the cost of the 3 billion cubic feet of gas that Cone Mills uses annually had increased 419 percent.

The aluminum industry is another that has been hard hit, both by recession and by escalating natural gas prices. Aluminum plants that use hydropower can produce a pound of product for 4 cents spent for energy, but some that have to use natural gas run into energy costs of 40 cents a pound. With the finished product priced at 76 cents a pound—and sometimes selling for 50 cents on the spot market—there is no margin left for profit. As a result, the Aluminum Company of America (Alcoa), the world's largest producer, and Reynolds Metals Company closed down two gas-powered plants along the Texas coast when power costs soared. Analysts estimated that their fuel costs were running about 30 cents a pound, too heavy a burden in a depressed market.

Not only are some businesses crippled by such natural gas increases, but the general productivity of American businesses is also affected, according to a study made by Dale W. Jorgenson, a Harvard University economics professor. By analyzing the performance of thirty-five industries since 1972, Jorgenson found that those most dependent on energy showed slower increases in productivity. They invested in energy-saving rather than labor-saving devices, they substituted labor for energy—and so suffered losses in productivity.

Such has been the effect of a "price creep" not half so severe as the accelerated decontrol the Reagan administration is engineering. Yet Big Oil still clamors for full decontrol. Exxon enclosed Big Oil's message in a little folder sent to its stockholders with their first dividend checks in 1982.

The Act [the Natural Gas Policy Act of 1978] was designed to move prices of "new gas" . . . from their cheap controlled levels to free market prices by 1985. This was to be done in gradual stages through a pricing formula based on the expectation that oil would be selling for $15 a barrel in 1985. Existing or "flowing" gas would remain under control.

What happened then was that oil prices doubled in 1979 and continued to rise. Already, the world oil price is near $35. So, instead of gas prices keeping pace with oil, they have fallen far behind. This discourages the production of natural gas and encourages consumption.

Exxon feels that decontrol would price natural gas at the true cost of alternative fuels, and so cut waste and encourage production.

Raymond Booth, assistant general manager of the natural gas department of Exxon in Houston, later expanded on this message. He argued that swifter decontrol would spare consumers a rude shock in 1985 when controls are to be lifted (So let's give them the shock now in the heart of the worst recession since 1932 and cash in on extra profits for the next three years?); he opposed any windfall profits tax designed in part to help the poor pay for higher heating costs (let the government take care of them); he contended decontrol would spur exploration and give us more gas (ignoring the fact that industry had been exploring to its limit in 1980 and 1981); and he considered it just and equitable that natural gas should be priced as high as heating oil by 1985 (so there would be no cost advantage in switching from heating oil to gas and Big Oil could continue to call the tune on *all* energy prices).*

To say that all this is pure sophistry is the kindest description that comes to mind. A product that costs one-third as much as oil to produce and market would be priced just as high as oil—and wouldn't that be a windfall for Big Oil? There would be no escape from ruinously high prices for the American homeowner. The heating oil price gouge would fade into insignificance compared to this new natural gas price gouge.

One of the phoniest arguments advanced by Exxon and Big Oil was the contention, endlessly repeated, that cheap

---

* Some natural gas companies were almost as opposed to deregulation as consumers. James T. Dolan, Jr., president of New Jersey Natural Gas, charged proponents of deregulation "are, primarily, the poor, misunderstood, low-profit, hard-pressed oil industry." Decontrol, he said, "would cause natural gas consumers to absorb, in a short period of time, price increases solely to satisfy the profit wants of the oil industry and to help lessen the price advantage of natural gas over fuel oil" (*Asbury Park Press*, November 15, 1981). Willis Strauss, board chairman of InterNorth, Incorporated, parent of Northern Natural Gas Company, declared that speeding gas decontrol could result in the loss of markets that would switch to cheaper fuels (as had happened in 1980) and result in a drop in demand that would stifle increased gas production. Industrial consumers, he said, "would either switch fuels or be driven out of business" and residential consumers solely dependent on natural gas "would be breaking down the doors of their political representatives demanding protection" (See *Oil & Gas Journal*, March 9, 1981).

natural gas prices were leading to waste and higher, needless consumption. Nothing could be more erroneous. Official D.O.E. figures showed that, despite hundreds of thousands of residential conversions to natural gas, actual consumption fell 7 percent in 1979 to 19.5 TCF—and it fell even more in 1980, to 18.7 TCF.

What was the explanation for this seemingly greater use and lower consumption? More efficient devices were part, but not all of the explanation. I had several long conversations in November 1980 with a spokesman in the office of the director of natural gas within D.O.E. The office acknowledged that there were abundant supplies of natural gas, that it was relatively cheap—and yet the market for it was soft. It seemed like a conundrum, yet there was an explanation for it. Not only were the large users of natural gas cutting consumption as much as possible by energy-saving programs, but many were switching to heavy residual oils. D.O.E. explained that the oil companies had been overloaded with these No. 5 and No. 6 oils that can be used to heat large commercial and industrial buildings. "In many cases," the spokesman explained, "it is easy for such large consumers to substitute oil for gas as simply as throwing a switch, and this is what has been happening, since some of these residual oils have been cheaper than gas."*

This experience indicates that decontrol and the higher prices to come with it would still further depress the natural gas market. It would leave the nation in the future, as it has been in the past, the helpless prisoner of whatever energy price benchmark OPEC and Big Oil see fit to set. One of the

* Herman Schwartz, a lawyer on Senator Metzenbaum's staff, told me the same thing when I interviewed him on March 18, 1981. Schwartz said there had been such a 1981 oil glut that No. 6 oil, the leftover substance from the refining process, had been selling very cheaply and undercutting natural gas. "So you have a situation," he said, "where natural gas supplies are up, demand is down—but do prices come down? No, they go up. It is madness. It all comes down to ideology and greed. There is nothing rational in the situation."

great fallacies of the 1978 natural gas act was that it accepted the Big Oil thesis that natural gas should be priced on a parity with heating oil. If this philosophy is to be etched in marble, only catastrophe lies ahead for the nation.

D.O.E., the lapdog of Big Oil, seems to accept this prospect with equanimity. In March 1982, D.O.E. looked into the future and told us what we might expect. It predicted that by 1995, without taking inflation into account, gasoline would be selling at $2.20 a gallon. With inflation figured in, the price would be $5.50 a gallon. And D.O.E., which was championing natural gas decontrol, forecast that the biggest boosts—and the worst shocks—would come from huge natural gas price increases.

Unless this trend is stopped—and with the Reagan administration in power and Congress almost moribund, there seems little hope of that—Big Oil will wind up owning the nation even more than it does today. Representative Dingell, in a hearing in Washington on March 12, 1981, described what was happening.

Direct and indirect energy costs have contributed about one-third to the total inflation in consumer prices in 1979 and 1980. Put another way, but for the increase in energy prices, the rate of inflation the last two years would have been about 8 percent, or roughly the same level as in 1978. . . . [Other estimates put energy's contribution to the inflation rate at 40 percent.]

According to a report by the Democratic Study Group, oil company profits accounted for a staggering 40 percent of all profits earned by all U.S. manufacturing firms in 1980. [*Business Week* cited the same figure]. In other words, some 30 oil companies made almost as much profit as the 9,000 other major U.S. companies. They did so at the expense of those 9,000 companies.

In a later report in December 1981 Michael Barrett, chief counsel to Dingell's committee, spelled out the economic damage done to the American industrial base by this lop-sided concentration of profits in the hands of Big Oil. He found that, since 1973, there had been a steady "redirection of economic strength into one industry—energy—at the expense of the rest of the industrial base. Capital that would have been available to rebuild American industrial machinery across the economy has been diverted into energy and especially into the oil and gas industry. . . ." This, Barrett wrote, "is a much more significant problem than government spending or other inflating activity."

Four of the five largest U.S. industrial companies were oil companies. Barrett added, "This shift into energy exploitation has reduced the U.S. to the status of a lesser developed country. We are a net exporter of food and raw materials, such as lumber and coal, and we are a net importer of value-added products such as machine tools."

Big Oil's greed, in other words, had reduced the greatest industrial power in the world to the colonial-like status of exporter of raw materials and importer of finished products.

14

The Spectre at Big

Oil's Feast

The date was June 1, 1981; the site a hearing room in the House of Representatives' Rayburn Office Building. Rep. Philip R. Sharp, chairman of the subcommittee on fossil and synthetic fuels, was making the first serious attempt by Congress to find out the truth about natural gas. The hearing room was packed with researchers, writers, lobbyists for energy causes and representatives of major companies. I sat at the end of a long reporter's table on the left-hand side of the room. Since my seat was closest to the congressmen conducting the hearing, a committee aide would bring me texts of each witness's prepared testimony. No sooner did I have the stack in my hands, trying to take one copy off the top and pass the rest to my colleagues in the press corps, than the pack charged. It was like being attacked by a swarm of two-legged barracudas—snatching, grasping hands tried to rip copies of the texts out of my hands before I could get my own copy or pass the rest along to the reporters nearest to me at the table.

This frenzied assault seemed like an indication of high interest in the subject, and I wondered afterwards what had happened to that interest, since precious little about the hearing found its way into print. Perhaps writers not already familiar with the subject assumed that D.O.E. was an impartial and authoritative voice, in which case confusion and uncertainty would be understandable.

I knew differently, and nothing that D.O.E. did could

surprise me. Besides, I had a special interest in this hearing. Having studied the possibilities of natural gas for more than a year, I had become convinced that a determination of how much natural gas we had, or could get, was fundamental to the development of any rational energy policy. There were questions that cried for answers. Did we have enough natural gas to last for centuries? Did we have enough just to heat our homes, cook our food and heat our hot water for the next fifty years? Or did we have such supersufficiency that we could convert vast amounts into liquid methanol—a motor fuel that could replace gasoline?

My attention had been directed to the possibility of the methanol alternative the previous November when I talked to Garvin McCurdy, a consultant of Carson Associates of Alexandria, Va., and Boston. McCurdy was an expert on methanol. "If the extremely large quantities of natural gas projected, not only by Paul Jones but by others, are correct, then we have no problem," he said. Automobile engines could be converted to run completely on methanol.

This was precisely what the Bank of America, one of the nation's largest banking institutions, had been doing on the West Coast. Having suffered from the 1979 phony gasoline shortage, Bank of America officials had decided they didn't want to be caught short like that again; and so they began to experiment with various methanol-gasoline blends. They tried using a 2 percent methanol blend, then 4 percent, 8 percent, 12 percent, and 18 percent.

"At 2- and 4-percent blends," McCurdy said, "methanol converts gasoline into a premium grade with little corrosive effects."

Such judicious use of methanol, he explained, decreases the need for an octane booster in unleaded gasoline because methanol itself is the booster. Bank of America, however, had not stopped at this point in its experiment with methanol. For almost a year, it had been running a portion of its car fleet on pure methanol. McCurdy himself had driven a methanol-powered car.

When you get behind the wheel of a methanol-fueled car, the main thing that impresses you is performance. It gives you a peppier engine. It burns absolutely clean and has a very high octane rating, which means you could build engines that would give higher performance. In a state of advanced technology, methanol can be made significantly more efficient than gasoline or diesel fuel. There's a quantum jump that methanol can take that diesel can't.

If you had it at a price competitive with petroleum, it should prove to be a superior motor fuel, giving extra miles to the gallon.

This was a thought to make Big Oil shudder. Since its primary financial stake is in petroleum, it could hardly be expected to welcome the idea of so potentially dangerous an intruder. Especially since methanol, which is now made solely from natural gas, can also be produced from garbage, from peat bogs, from boring into coal mines to siphon off methane, from converting wood into methane. The possibilities are almost endless.

Naturally, there are new problems. McCurdy conceded that pure methanol is more corrosive in an automobile engine than gasoline. Engines would have to be modified— made of different metals. There would be similar problems with tank storage. And, of course, there would be problems of distribution, setting up sufficient stations to dispense methanol. All of this might entail massive shifts in the American economy. "We just have to get experience under our belt," McCurdy said. He added that he thought the role of government should be to determine what resources were available, what was the best way to go. Representative Sharp's hearings (there was to be another on June 9) were a first step in that direction.

The leadoff witnesses were from D.O.E.—Albert H. Linden, Jr., acting administrator of the Energy Information

Administration, and Kenneth A. Williams, director of the Office of Pipeline and Producer Regulation of the Federal Energy Regulatory Commission. Both could hardly wait to poor-mouth the possibilities of natural gas; they were, indeed, frankly contemptuous of it.

Linden saw only "a modest increase in drilling" (despite the frenzied activity the last two years): only such a gradual increase in reserves that it would "not replace production, therefore giving a yearly decline in reserves." Williams echoed the same line. He said that "although reserve additions have been increasing, annual production continues to exceed those additions and remaining proven reserves continue to decline." It was essentially the Big Oil–D.O.E. party line laid down in concrete during the Carter administration: We had only limited natural gas supplies and natural gas could be disregarded as an important energy resource.

Under questioning by committee members, both Linden and Williams admitted that current demand for natural gas was "soft" because "demand is constrained by prices." Then, when asked, "What can Congress do?" Williams snapped, "Price incentives. Deregulate." Having offered a solution that flew in the face of the problem, Williams and Linden continued to denigrate the natural gas alternative.

Congressmen pressed them further. Since the natural gas market was already "soft" at current prices, would the demand for natural gas be there at higher prices?

LINDEN: We see coal on a BTU equivalent as being twice as cheap as gas.

WILLIAMS: We expect demand to decline as prices continue to rise. With rising prices, people are going to find other ways to heat instead of using gas.

This back-of-the-hand, expert testimony must have been cheering news to Big Oil which owns the preponderance of the coal reserves of the nation just as it does petroleum reserves. Congressmen were obviously skeptical about the answers they were getting. They asked Linden the key question: "What if we decided to use more natural gas to reduce imports of foreign oil?" This elementary idea was obviously new to Linden. "We can make a study of it," he replied lamely, a concession that no study of the obvious had yet been made by D.O.E.

Rep. James M. Collins questioned the witnesses. Was there any danger of more natural gas shortages like those in the bitter winter of 1977? Linden was still the pessimist where natural gas was concerned. No, he conceded, there was no danger of such a shortage at the time, "but if we had a continuous cold winter it could be." It was testimony that ignored all the evidence about the enormous new quantities of natural gas coming on line, and then, of course, in 1982, the nation got Linden's "continuous cold winter"—one of the most brutal of the century—and there were no natural gas shortages.

Collins had talked to Bob Hefner and had obviously been impressed by the Anadarko driller's assertion that enormous amounts of gas can be produced from depths below 15,000 feet. Hefner had estimated, Collins said, that the Anadarko alone contains between 60 and 100 trillion cubic feet of gas. "And that's in just this one basin," Collins emphasized. "Yes, we are seeing a lot of reevaluations," Linden responded weakly.

In bold contrast to the fumbling testimony of the D.O.E. witnesses was the statement of George H. Lawrence, president of the American Gas Association (A.G.A.). "Resource base estimates for U.S. conventional recoverable natural gas," he told the committee, "range generally from 700 to 1,100 TCF, some thirty-five to fifty-five times current production levels. The more recent and well-known estimates are in the higher range, with older, less reliable estimates in the lower range."

Contradicting the testimony of the D.O.E. witnesses about dwindling supplies, Lawrence said

Seismic crew activity has reached the highest level in over twenty years, the rig count and gas well completions have risen to all-time highs, exploration and drilling activity in general has increased . . . and recent lower-forty-eight state reserve additions have reached the highest level in twelve years.

The principal obstacle to further exploration and development of natural gas supplies was a weak consumers' market, Lawrence said. Though some 400,000 American homes had been converted from oil to natural gas for heating in both 1979 and 1980, demand remained down. Lawrence urged the repeal of the Power Plant and Industrial Fuel Use Act of 1978 which banned the use of natural gas by public utilities after 1990 and curbed the use of gas by commercial and industrial establishments. These restrictive measures stemmed from the Carter administration's exaggerated assessment of energy shortages. Lawrence saw "an urgent need to repeal unreasonable natural-gas demand restraints" so that utilities and commercial and industrial consumers could make fuller use of natural gas supplies. Unlike Big Oil and the D.O.E. spokesmen, however, he strongly opposed any immediate decontrol of gas prices. He pointed out that, under the gradual increase permitted by the 1978 natural gas act, gas prices had been rising at a rate of 4 percent above that of inflation. Urging more use of natural gas and a continuation of the 1978 controls, he said

The immediate removal of most restraints of gas use by industry and electric power plants should result in gas further backing-out oil imports. This policy would soften an already soft world oil market, resulting in slower oil price increases. Finally, immediate, complete deregulation of natural gas would shock the domestic gas market because most gas purchase contracts contain terms that require gas prices to escalate suddenly and rapidly upon deregulation—to prices well in excess of free market levels. . . . Indeed, A.G.A. estimates that gas prices will double to all classes of consumers if all wellhead prices were immediately deregulated. This shock will, therefore, reduce gas sales and increase oil use, which will in turn cause higher world oil prices and domestic gas prices. . . . A.G.A. analysis shows that total, immediate deregulation would cause the loss of 1.9 TCF of gas load, increasing oil imports by 800,000 barrels/day.

Lawrence continued by explaining that natural gas has many uses besides heating and that supplies are adequate to take full advantage of them. Gas air-conditioning could reduce electric utility loads and obviate the necessity for building more generating capacity. Gas-fired cogeneration, using steam for both heating and electrical generation, "has been shown to be economically and environmentally beneficial," he said. Then he suggested another alternative—the use of natural gas in motor vehicles.

The use of compressed natural gas [CNG] in automobiles seems well justified on both a cost and an environmental basis. The principal market for CNG vehicles will be the urban automobile fleets. Methane—either natural gas or synthetic gas—has three major advantages over gasoline and diesel fuel as an alternative fuel: (a) It is more economical, (b) it is cleaner burning and (c) it is principally domestic. The primary disadvantages of methane in transportation (which can be overcome) are range and refueling limitations.

In summation, Lawrence said that, out of a total of 8.4 million barrels of crude oil consumed per day in nontransportation uses in 1979, "about 5.5 million barrels of oil equivalent/day could technically have been replaced by gas."

Throughout his testimony, Lawrence made it clear that he was relying solely on supply estimates from conventional sources in the lower forty-eight states when he concluded that the nation probably had available 1,100 TCF—enough at the rate of consumption when he testified (it is lower now) to last the nation for some fifty-five years. In addition to the conventional base, however, he emphasized that there were four other sources of "significant potential" in the lower forty-eight: (1) tight sands in the Rockies, northern Great Plains and the Southwest with trapped resources ranging from 400 to 1,000 TCF, (2) Devonian shales in the Appalachian, Michigan and Illinois basins with an estimated 200 to 1,200 TCF, (3) methane from coal beds ranging in amount from 200 to 500 TCF and (4) geopressured aquifers in the Gulf Coast region whose volume appears to be huge but is impossible to estimate.

In addition, there were other important resources: the 26 trillion cubic feet in the Prudhoe Bay field waiting to flow south as soon as the Alaska gas pipeline could be built; enormous supplies in new fields constantly opening in Canada and Mexico, both more available to the United States than Middle East oil in a crisis; and synthetic fuels from coal gasification and liquefaction. Putting it all together, it seemed from the testimony of this most conservative industry spokesman that reserve-resource base is genuinely stupendous, capable of lasting the country well into the next century even if fuller use is made of our natural gas supplies.

Other experts supported Lawrence and made D.O.E.'s gloom look ridiculous. H. E. (Gene) Wright, president of the Independent Petroleum Dealers Association of America, said that the Carter administration's contention that we were running out of natural gas had been "totally defeated." He testified that 3,850 rigs, a record number, were then drilling and that "only 3 percent or so of sediments have been tested so far."

Harry Kent, director of the authoritative Potential Gas Agency of the Colorado School of Mines, estimated U.S. total resources and proven reserves at 1,100 TCF—"on the order of twice all U.S. production to date." Dr. Charles Mankin, director of the Oklahoma Geological Survey, ticked off the huge gas basins that were just being brought into production all over the country and said, "I guess my view would be that the figures presented here would be some sort of minimal figures for supply."

One of the strongest statements made at the two hearings came from Henry B. Taliaferro, Jr., vice president of Bob Hefner's GHK Company. Taliaferro quoted statistics that made one wonder how D.O.E. witnesses could have concluded that reserves were being depleted and natural gas could be dismissed as relatively inconsequential.

Linden and Williams had contended that most of the new drilling was concentrated in fields already discovered and that wildcatting activity to find new reservoirs was minimal. Not so, according to Taliaferro. In 1981, $2 billion was being invested in deep-drilling in the Anadarko alone, most of it by wildcatters who were successful in bringing in 60 percent of the wells they drilled. He testified, "With less than 3 percent of the deep sediments explored prior to 1979, nearly 4 trillion cubic feet of proven reserves have already been established."

Turning to the Rocky Mountain Overthrust Belt, he declared, "Ten point three trillion feet of proven reserves have been discovered in the Overthrust as a result of drilling activity initiated in just the past few years." Potential reserves through the year 2000 are "estimated at 200 trillion cubic feet" in the Overthrust alone, he said—and only 5 percent of the sediments had been explored.

He cited in addition Amoco's huge finds in the deep Tuscaloosa Trend, two deep natural gas finds in the Appalachian Basin and stepped-up drilling in Ohio and West Virginia that had tripled the activity in those states between 1975 and 1980.

Unlike oil, Taliaferro said, great amounts of natural gas are trapped in deep geologic formations "that underlie vast portions of the lower forty-eight United States." There are, he said, "thousands of cubic miles of deeper sediments where the search for nonassociated natural gas has just begun. . . . Today, wells being completed at [such] depths are among the most prolific of any in the United States, with reserve additions and deliverability at ten times the national average."

Taliaferro summed up. "The choice for Americans today is domestic gas or foreign oil. We can curtail natural gas markets and rely on higher oil imports. Or we can develop our natural gas supplies. These supplies are cleaner burning and sufficient to displace nearly all foreign oil."

Similarly graphic testimony was given by another producer with long firsthand experience in the field—Thomas J. Vessels, executive vice president of the Vessels Oil & Gas Company of Denver, Colo. Vessels testified that his company had drilled for oil and gas in Montana, South and North Dakota, Wyoming, Nebraska, California, Oklahoma, Texas, Colorado, Kansas and Saskatchewan, Canada. It had brought in so much gas that it couldn't sell all of it in the "soft" market then existing.

"We have eight wells shut down [in the Rocky Mountain area]," Vessels testified. "Some operators in Ohio and Pennsylvania have restrictions up to three months in delivering their gas through lack of demand on the pipelines." He said his firm had forty other wells that could be profitably developed, but the lack of demand had prevented the investment of capital necessary to lay connecting pipelines.

Vessels was operating two hundred wells in the Denver-Julesburg Basin in southwestern Colorado. He described his firm's experience in drilling a sixty-three-well wildcat farmout from Amoco, beginning in 1970. This drilling developed the Wattenberg gas field and the Spindle, Surrey, Singletree and Peoria oil fields. Vessels explained that the Wattenberg field had been drilled originally to a depth of 8,000 feet. In going that deep, drillers went through the Spindle Oil Field existing at 5,000 feet. The Spindle, he said, might never have been discovered except for the deeper drilling for gas. The result: In 1980, the Wattenberg field produced 29 billion cubic feet of gas, and the Spindle field, in addition to its oil production, produced another 20 billion cubic feet of what Vessels called "casing head gas." He pointed out that the Wattenberg drilling had "resulted in the discovery of numerous oil and gas fields, which would never have taken place had the original gas wells not been drilled." This experience, he said, led him to believe that many shallow wells producing gas from depths of only 2,000 to 3,000 feet were just tapping the top of larger reservoirs that lay buried under thick layers of deep sediments.

The cumulative testimony was overwhelming. It became obvious that the problem involved, not the supplies of natural gas (there was an abundance), but the nation's failure to make use of the supplies at hand. In this perspective, the 1978 Power Plant and Industrial Fuel Use Act made little sense. Rep. W. J. (Billy) Tauzin pointed out that the act required gas-fired utilities to stop using gas and switch to coal by 1990. James Collins, the Texas Republican, added that coal requires an entirely different type of plant; the changeover cannot be made as easily as it can from oil to gas or gas to oil. This meant that utilities would soon be facing a deadline for the rebuilding of their plants to meet the 1990 cutoff date. The cost would come to some $50 billion, and utility costs would soar.

(There was another important factor not mentioned at the hearing—the risk of further ecological damage. The cluster of coal-fired utility plants in Ohio and Illinois was already spewing tons of sulphur emissions into the higher atmosphere. This was coming down as acid rain. With the west-to-east wind flow, this falling poison from the sky was spreading a deadly trail across Canada and the northeastern states. Canada had made angry protests because the acid rain had killed every living organism in many of its streams and lakes. The beautiful lakes in New York's Adirondacks region had become similar stretches of dead water. Even the facades of public buildings in New York City were being eroded. If more utilities were forced to shift to coal, this environmental damage would be multiplied many times over unless costly and efficient scrubbers were installed in stacks to filter out the emissions. Given the Reagan administration's attitude that damage to the environment didn't matter compared to costs to business, the likelihood of the installation of efficient control devices appeared dim.)

There was general agreement among both committee members and witnesses that the 1990 ban on utilities' use of natural gas should be repealed, but there was no such understanding about how natural gas should be priced. Rep. Tom Corcoran mentioned that, when the 1978 Natural Gas Policy Act was passed providing for oil-gas parity by 1985, "it was assumed that crude oil prices would be $15 a barrel." This brought an outburst of spontaneous laughter in the hearing room. "Unless we act to revise the schedule now," Corcoran continued, "it will be almost irresistible in 1984 for Congress to impose controls again." This, apparently, was such an appalling thought that Corcoran's remark sank into a smothering silence.

In questioning Taliaferro, Rep. Mike Synar tied together the two issues of supply and price. First, he asked Taliaferro, citing his testimony about abundant supplies, "Then the official figures don't present an accurate picture?" Taliaferro said that they didn't. Next, Synar asked Taliaferro, "How can you charge $7 per thousand cubic feet if there is a surplus?" (Actually, this figure was too low.) Taliaferro answered, "If artificial demand restraints were removed now, the market will clear and prices will moderate substantially." Synar asked, "If you are right, how long will it take for the market to settle down?" "1985," Taliaferro replied. He advocated repeal of the power and fuel use act to broaden the consuming market.

In the aftermath of the hearings, Congress took one positive step. On July 29, 1981, the House and Senate agreed on what is known as the Omnibus Reconciliation Act of

1981. The major change was the repeal of the 1978 power plant provision banning the use of gas by utilities for electric generation after 1990. Spokesmen for utilities in California, where gas helps them comply with the Clean Air Act, and for utilities in the southwestern gas-producing states were instrumental in getting the power plant ban repealed. The restrictions on further commercial and industrial use had already been abandoned.

While the lifting of restraints on natural gas usage could in theory increase consumption, the Reagan administration's back-door decontrol maneuvers and its champing at the bit for full decontrol posed a price threat to that very increased usage. In 1981 and 1982, there was suddenly a huge world oil glut, largely due to the recession and decreased consumption as well as more drilling; in these circumstances, as my D.O.E. source had acknowledged, natural gas was not price-competitive with the cheaper forms of heavy residual oils. Escalating prices under accelerated or full decontrol could only compound the problem.

It became ever more obvious that natural gas prices should never have been tied to petroleum prices. The two commodities are entirely different, and, as Edwin Rothschild had pointed out, natural gas costs about one-third as much to produce and distribute. Gas requires no twelve-hundred-foot-long tankers costing billions of dollars to build and sail across the oceans, the nation is already threaded with more than a million miles of pipelines, mains are in place in major cities and towns and connecting lines to utilities and factories that were threatened by a natural gas ban haven't been removed. Most important, natural gas does not require huge and costly refineries; unlike oil, it does not have to be broken down and processed into varying grades and products. Basically, all it needs is to be stripped of its moisture and compressed. Minor expenses involve the maintenance of pressure to insure the continuous flow of gas through pipelines and mains and the creation of storage facilities to keep gas available for winter consumption during the lower-usage periods of summer. So natural gas pricing parity with oil that Big Oil clamors for can benefit only Big Oil.

The Big Oil cartel has jerked the nation around almost at will. It virtually owns the government. As the aide of one frustrated congressman said to me, "The Democratic party is just as much a Big Oil party as the Republican." The result is that a relative handful of major companies has locked up the nation's oil and coal resources and created a high price structure beneficial only to themselves. The only threat on their horizon is possible competition from cheaper natural gas. And so the remaining question is can they control this, too?

The answer seems to be Yes, of course they can. Control of natural gas may not come as easily because there are more independents in the natural gas business. Nevertheless, the majors have been buying up the mineral rights to huge land tracts, and they are drilling at a faster pace, bringing in tremendous flows of gas from blockbuster wells. They control more than 50 percent of natural gas production and 60 percent of pipeline distribution. This means they have enormous leverage, enormous clout. And there is evidence they are using it.

When I talked to David Schwartz, the former Federal Power Commission expert, he had just returned from Dallas, where he had met with a number of oil men. He told me:

I talked with them about what is happening in the San Juan Basin and the Permian Basin, and it became clear that what is happening is that they are holding back on "old," regulated gas. They are selling only deregulated gas. That is price-fixing, pure and simple.

You do not have a competitive market. You have a monopolistic, oligopolistic market. New contracts are concentrated in the hands of eight firms, and they control above 50 percent of the market. There are all kinds of interlocks—bank interlocks, pipeline interlocks, interlocks with independents. Producers don't assume all the risks themselves. They form joint ventures. A major will combine with an independent, letting the independent do the drilling. If they strike it rich, the independent gets perhaps one-eighth of the profit. If the well is dry, the independent is the principal loser. The big majors dominate the market through these interlocks and joint ventures. What this represents is a major transfer of wealth from the citizens of this country to the major companies and the landowners.

At the time we talked, Big Oil had not yet pulled off one of its most astounding coups, one that demonstrated the might of its muscle and its influence over government—the now notorious Alaskan gas pipeline deal.

15

From Zero to Billions

I have this marvelous dream. I can just see how I am going to become a millionaire—possibly even a billionaire like the late H. L. Hunt. But I'm not a greedy man; I don't worry about the billions—just a million or two will do. I feel almost certain of getting those millions; this is a fool-proof scheme. There's just one hitch to it. I need $40 billion to get started. Now, of course, I don't have $40 billion, and, of course, the banks won't lend me $40 billion. So what can I do to make my beautiful dream become a reality? Well, I know a lot of congressmen and senators. I'm sure they would be just dying to help. All I have to do is push the right buttons—and Congress will guarantee the $40 billion for me. That is, Congress will pass an act that will let me go ahead with my dream plan and pass all the costs along to the American people. It won't cost me a cent—I won't have to risk a cent. And the returns will be fabulous. Perhaps I'll even get a billion or two after all.

Sound crazy? Well, nothing is too farfetched in the world of Big Oil. The fantasy I've just described is a reality. In simplified terms, this is precisely the kind of sweetheart Christmas present that the Congress of the United States bestowed upon Big Oil when it passed the Alaska Natural Gas Transportation System Act just before it ducked for home for the holidays in December 1981.

The deed was accomplished over the vociferous protests

of consumer groups and with the aid of all-out lobbying by some of the most powerful politicos of both parties. Capitol Hill veterans who considered themselves inured to lobbying excesses were shocked by the massive pressure applied for the passage of the pipeline bill. The result was all but inevitable: The consumers lost—again.

The story of the Alaska gas pipeline goes back to 1968 when oil and natural gas in huge quantities were discovered in Prudhoe Bay on Alaska's North Slope. Since then, development of the Prudhoe Bay reservoir has established that it holds 26 trillion cubic feet of gas and the American Gas Association, for one, believes that Alaska has the potential of ultimately producing 177 TCF. This is an enormous resource, but one with obvious problems: How is this gas to be moved over the 4,800 miles that separates it from consumers in the lower forty-eight states? And at what cost?

Several proposals were made over the years, some of which ran into Canadian opposition. Finally, in September 1977 the Canadian and U.S. governments agreed on the route the proposed pipeline would take. It would run south through Alaska, curving slightly to the east, then cross 1,347 miles of Canadian territory to Calgary. At Calgary, the pipeline would split into two sections. The eastern stem would slant diagonally across Canada and the northern states to Chicago, a distance of 1,117 miles. The western stem would run another 281 miles to the Canadian border and then 911 miles through the western states to a terminal at San Francisco. At the western and eastern terminals, the pipeline would feed into already available transcontinental systems, making Alaskan gas available to all of the lower forty-eight states except Vermont.

This Alaskan pipeline system was envisioned as the most gargantuan task ever undertaken by private enterprise. And, indeed, it would have been had it been tackled by private enterprise alone. From the first, however, there had been serious doubts about the ability of private enterprise to finance and carry to completion such a mammoth project. Some of the early bidders had taken the position that it couldn't be done without government assistance or government price guarantees for the final, delivered product.

At this point, a tough-as-nails onetime oil field roustabout who had fought his way to millionaire's status entered the picture. John G. McMillian, head of Northwest Energy Company, a pipeline consortium based in Salt Lake City, Utah, had been a longtime, heavy contributor to Democratic campaigns—a record that certainly was no handicap when he sought the Carter administration's approval of his pipeline plan. In September 1977, he succeeded where others had failed, obtaining Carter's backing for his proposal to build the pipeline at an estimated cost of $10 billion.

Carter's *Decision and Report to Congress on the Alaskan Natural Gas Transportation System* spelled out these conditions in recommending the award to McMillian:

- The line was to be privately financed, with no government guarantees or consumer risk-bearing.
- The Big Oil North Slope producers—Exxon, Atlantic Richfield and Standard of Ohio—were to have no management or ownership role in the pipeline.
- The three Big Oil companies would build the gas conditioning plant and absorb the cost.
- The pipeline would be granted an incentive rate of return. This was later set by the Federal Energy Regulatory Commission at 17.5 percent above actual costs.

On the same day that President Carter sent this proposal to Congress, McMillian told two subcommittees of the House of Representatives

The president's decision requires the Alcan project to be privately financed in its entirety. The United States and the Canadian governments will not be called upon for financial guarantees. *Nor will the consumer have to bear the hypothetical burden of the noncompletion of the project.* Instead, other primary beneficiaries of the project will be called upon to provide the necessary financial backing. We believe that Alcan can obtain the necessary financing from Canadian and United States sources. [Italics added.]

This turned out to be a free-enterprise pipe dream. Four of the nation's banking powerhouses—Chase Manhattan, Citibank, Morgan Guaranty and Bank of America—looked at McMillian's project and didn't like what they saw. One could hardly blame them unless one expected them to be altruists. Inflation was sending construction costs to astronomical levels, and there was great uncertainty about the energy market. More natural gas discoveries were being made in the lower forty-eight states; huge and much closer reservoirs in Mexico and Canada were being developed. Gas from these sources should be cheaper than gas that had to be transported 4,800 miles from Prudhoe Bay. These circumstances gave conservative bankers reason to pause and ask themselves a serious question: With the almost day-by-day burgeoning cost of the pipeline project and the improving prospects for other supplies, would it be possible to find a market for the Alaskan gas after we got it? The bankers had their doubts, and doubtful bankers don't risk money.

Unable to get the financing he had so blithely promised, McMillian still struggled to turn his pipeline dream into reality. By the time four years had passed, cost estimates for the project had mushroomed from $10 billion to $40 billion, and there was no guarantee that it could be completed even at this price. If bankers hadn't liked the deal originally, there was no chance they were going to love it at these astronomical figures.

Enter the Reagan administration. Enter powerful pols of both parties. In 1981, a new McMillian proposal was drafted for the approval of Congress. It wiped out every safeguard for the consuming public that had been included in the original plan approved by Carter. The revised scheme gave Big Oil the brass ring trimmed with gold. It provided:

- Big Oil—Exxon, Arco and Sohio—would become 30-percent owners of the pipeline.
- A gas processing plant in Alaska, estimated to cost $4 billion—and not just to process gas for consumers, but to produce propane, butane and ethane for the oil companies—would *not* be financed by Big Oil but would be "rolled in" to the pipeline's overall cost.
- Even before the pipeline was completed, even if it never was, potential consumers were to be billed in advance and forced to pay the construction costs of each finished segment. (It wasn't clear just what would be considered a "segment." What was clear was that consumers were going to bear the financial burden.)
- Worse still, if that is possible, consumers were to have no recourse in the future, no matter how high their bills might be. The banks, it was said, would not participate without an ironclad guarantee that consumers would have to pay whatever the price was—and so one of the waivers in the deal barred any future consumer appeals to the Federal Energy Regulatory Commission for rate adjustments.

A staff memorandum prepared for Rep. Philip Sharp on June 24, 1981, put the issues into sharp focus.

The significance of allowing commencement of billing for sub-segments of the project before completion should not be underestimated. In essence, it would provide a consumer guarantee that the construction investment would be repaid even if the pipeline is never useful. It would take almost all the risk out of the loans to be made to the project, passing that risk to the consumers. It would allow the project to avoid a crucial test of its worth—the test of private financeability—and would represent a decision that the project should be completed regardless of cost because consumers would have no way of limiting their contributions. . . . Although the risks of construction are passed to the consumers, the reward for bearing the risk appears to be left under these waiver proposals with the sponsors—their rates and return formulas are to be changed only at their requests.

How did such a lopsided and patently unfair piece of legislation ever get drafted? How, once drafted, did it ever get through Congress?

The answers are to be found in the super high pressure lobbying that surpassed virtually anything Washington had seen before. It was an act that may stand for a long time as the premier, if not shining example of the cynical aphorism that lobbying is the fourth arm of government.

To begin with, the waiver package that abrogated all consumer rights and gave the moon and the stars to Big Oil and the banks was drafted by Rush Moody, a former Federal Power Commissioner. Moody is a law partner in the firm of Akin, Gump, Strauss, Hauer and Field, of Houston. The Strauss stands for Robert S. Strauss, the former national chairman of the Democratic party during the regime of Jimmy Carter. Once drafted, the pipeline bill was pushed by a host of other political brokers. A firm in the thick of the lobbying was White, Fine and Verville. The White stands for Lee White, who was chairman of the Federal Power Commission under Lyndon Johnson. White said his colleague, John Atkisson, was handling the firm's pipeline work on the Hill.

Anne Wexler, a former aide to President Carter, was drawn into the action. So were law firms associated with the present Democratic National Committee Chairman Charles T. Manatt, and with yet another former Federal Power Commissioner, Don Smith.

This array of prominent Democratic figures and former officials with contacts on the Hill was supplemented by an equally massive rallying of Republican stalwarts. The Reagan administration seemed at first lukewarm to the pipeline proposal. To change this attitude, McMillian engaged Peter Hannaford, a former speechwriter for Reagan and a partner of top White House aide Michael Deaver. Hannaford, it just so happened, had also purchased the public relations business of deposed National Security Adviser Richard V. Allen. Jack Ferguson & Associates were hired to help Hannaford in lining up G.O.P. support.

The Senate was soon in hand despite the opposition of some dissidents like Senator Metzenbaum and Sen. Paul E. Tsongas. The House, however, was a more difficult matter. Congressmen like Dingell, Moffett, Gore, Ottinger, and Edward J. Markey opposed the project as far too expensive. They argued that it would make the cost of gas to consumers exorbitant and that it would draw away capital funds that would be much better spent on the modernization of our industrial plant.

This opposition spurred an intensified lobbying effort. More than a dozen prestigious law firms became involved. The Big Oil companies had their own lobbyists; so did all eight firms in McMillian's consortium; so did the big steel corporations and firms like Westinghouse, potential suppliers of pipeline equipment and controls.

Behind it all was McMillian himself. He relied heavily on his ties with prominent Democrats. He contributed $10,000 for a table at a fund-raising dinner sponsored by the Democratic Congressional Campaign Committee. By an odd coincidence, Rep. Tony Coelho, of California, chairman of the committee, later became one of the pipeline's most ardent supporters. Then McMillian donated $5,000 to former Vice-president Mondale's Committee for a Future America, another 1982 congressional fund-raising operation. On the same day that he made this contribution, McMillian named Mondale himself a consultant to the board of Northwest Energy Company at an unspecified retainer.

Mondale's connection with McMillian's power play raised a lot of eyebrows in Washington, especially in view of the former vice-president's most poorly kept secret—his intention to seek the Democratic presidential nomination in 1984.

The eyebrows really shot higher when it was disclosed that Mondale had made a personal telephone call to Representative Sharp. Nobody would say what they discussed, but a Mondale spokesman denied that the former vice-president had been lobbying for Northwest Energy. The explanation was that he had just called Sharp to find out when the congressman was going to hold one of his hearings on the pipeline proposal. It was an explanation received with considerable skepticism in a Washington awash with lobbyists.

Against this array of political muscle, consumer advocates fought a valiant but doomed battle. The California Public Utility Commission (PUC) expressed its opposition in a letter to Sharp. It made these points: If the pipeline should be built for national security reasons, then the whole nation "should share in the risk of the noncompletion of the project"; waiver of the rights of consumers to seek fair rates through appeals was "a violation of the due process rights of those who pay these expenses"; if the gas conditioning plant to produce profits for the oil companies was to be included in the package, then the consumers who were the real investors should receive "a more equitable share of the natural gas liquids extracted from the plant;" and, finally, the California PUC doubted that the Prudhoe Bay gas, estimated to sell for between $15 and $20 per thousand cubic feet if the pipeline was completed by 1987, could be marketable at such an extravagant price.

The Colorado Public Utility Commission also opposed the contemplated rip-off of consumers on similar grounds. It argued that, if prebilling were to be permitted, then consumers should be credited with the advance amounts they had paid, plus interest, when the gas began to flow. The Colorado PUC was especially incensed at the waiver of the ratepayers' rights of appeal, contending this "could remove any protection the ratepayers may have against paying unjust and unreasonable rates for Alaska natural gas."

Sharp received an emphatic and hostile legal opinion from Edward Petrini, attorney for the National Consumer Law Center. Petrini pointed out that this was the first time an attempt had been made to force consumers to pay for labor and material costs during construction. Electric utilities, he said, had never tried to recover such costs "prior to the plant actually coming on line." He added, "By permitting noncapital costs of construction, as well as the capital costs of construction, to be recovered early, adoption of the waiver proposal would mark an important and, as far I know, unprecedented departure from conventional rate-making principles."

Petrini made the additional point that prepayment of such costs, no matter how high they might be, would remove the major incentive to complete construction on time. "Utilities have a stronger incentive to complete capital construction projects quickly if they do not begin recovering costs on their investment until the project is completed," he wrote. "By permitting early recovery of costs, this incentive is diluted."

Consumer advocates like Edwin Rothschild opposed forcing consumers to pay for the $4-billion conditioning plant when the plant was going to produce petrochemical feedstocks that command high prices at no capital cost to Big Oil. "If consumers are going to be forced to pay for something producers should rightfully be paying for, then consumers should at least receive compensation in return," Rothschild argued. "In other words, the value of the plant liquids should be used to offset the rates charged to consumers for the conditioning plant."

The waiver of the consumers' rights to appeal rate charges would "set in concrete the rates they would have to pay, regardless of future developments," Rothschild pointed out. Suppose the sponsors assumed for depreciation purposes that the life of the reserves would be twenty years, but instead lasted for forty. The producers would have a free ride for the extra twenty years while the consumers would get no adjustment in their rates. Or suppose, Rothschild said, that technological advances such as improved compressors should reduce operation and maintenance costs. Again, the entire benefit would go to Big Oil—the consuming public would continue to pay at the previous high rates "set in concrete."

The Reagan administration, composed of free-enterprise ideologues, was removing virtually all free-enterprise risk from Big Oil and the banks and loading it on the backs of consumers, Rothschild continued. "I do not know what you call this arrangement," he told Sharp's committee, "but it certainly is not free enterprise and it is not private financing."

He predicted that, once the pipeline bill was passed, the sponsors would come back for an additional meal at the federal trough. "Each of the sponsors was asked if the waiver package was sufficient to insure construction and completion [of the project]," Rothschild said, "and none of them would state that it was. . . . Robert H. Graham of Citibank isn't sure whether the package 'will be sufficient,' while Stephen W. Jenks of Morgan Guaranty Trust states, 'Whether or not this package will be sufficient to ensure such financing we are unable to say at this time.' "

In conclusion, Rothschild made a suggestion that must have turned Big Oil moguls pallid as any ghost. He said that perhaps it would make more sense to turn the natural gas into liquid methanol which could be sent down the already existing Alaskan pipeline—a proposal that would turn Big Oil's gas into a competitor with Big Oil's own petroleum for the American automotive market. This was an abomination not even to be considered in a world in which Big Oil's multibillion-dollar profit structure must be secured and maintained at whatever costs to the public good.

Undeterred by all such commonsense arguments, pressured on every side by the unprecedented swarm of lobbyists,

McMillian's boondoggle, as Rothschild dubbed it, whipped through Congress unchanged. On November 19, the Senate passed the bill by the lopsided vote of 75 to 19, and sent it on to the House.

There the measure faced stiffer opposition. Though Representative Sharp, Mondale's telephone mate, had decided to sponsor the measure despite all the adverse reports he had received, a solid bloc of less committed congressmen waged a bitter battle against it. Two votes were held. On the first, on December 9, 1981, the pipeline measure passed by a vote of 233 to 173, much closer than had been anticipated. Opponents charged, however, that parliamentary procedures had been violated and that the vote was illegal. The House Democratic leadership responded by calling for a second vote the next day, December 10, and this time the measure passed 230 to 188.

The action left an aftermath of bitterness and uncertainty. "This is nothing more than a subsidy for the oil companies and the banks," Representative Markey charged. Rothschild said, "Congressional Scrooges today voted in favor of a Christmas windfall for the pipeline's sponsors and paid for it with an extra tax on consumers." Ralph Nader, the consumer activist, accused Speaker Thomas P. O'Neill, Jr., of betraying a pledge made to consumer groups to delay the vote. "The momentum was on our side," Nader said. "Another week and I think we would have won."

The legality of the pipeline legislation was promptly challenged in federal court in the District of Columbia. In a highly unusual action, twenty-four members of Congress, representatives from five states and several consumer groups filed suit, seeking to get the legislation outlawed. This made bankers even more skittish. Even with the no-risk guarantees written into the act, the bankers still weren't certain that they wanted to commit their money. One banker told *The Wall Street Journal* that the sponsors were $9 billion short of raising the $27 billion they needed just to build the

Alaskan segment of the pipeline. Another said there is a "big, gaping hole" in the financing of the project.

A major reason for the bankers' hesitation was their doubt that the Alaskan gas, even if it were to be delivered by 1987, could be sold in the lower forty-eight states where supplies were plentiful and demand was soft. Rothschild was predicting that the pipeline bill would add another $150 a year to every homeowner's bill on top of the already rising prices. The average industrial user in Ohio might have to pay an additional $41,000. In such circumstances, industrial users would certainly switch to residual oils as they could easily do—and, indeed, had already been doing. Even some homeowners might convert back to oil again, exacerbating the entire energy problem.

It was, to put it mildly, a mess. Bankers said it would probably take six months for them to complete the economic and engineering studies they needed before they could make a decision. The best estimate was that it would be the end of 1982 before serious negotiations to put together a financing package could even be begun.*

* Even this initial estimate proved too optimistic. On April 30, 1982, the pipeline sponsors met in Salt Lake City and announced that they were putting off development of the pipeline for two more years. They had not been able to raise a needed $20 billion in private financing as a result of the "current excess world energy supply, depressed crude oil prices, lower levels of economic activity in the United States and abroad, and uncertainties in the financial markets." Also, the estimated cost of the project had escalated to $43 billion. With the likelihood of continued inflation, continued high interest rates and rising construction costs, some opponents of the pipeline like Edwin Rothschild believed that the death knell might have been sounded for the project. They thought that some alternative methods would have to be used to make Alaskan gas available. Two were being suggested. Arco, one of the principal Prudhoe Bay developers, favored building a methanol plant in Alaska and sending gas in liquid form through existing pipelines (the same proposal Rothschild had made in his testimony before the Sharp committee in the House). Another alternative was to liquify the gas in Alaska (LNG) and to ship it by LNG-equipped tankers to the West Coast, where it could be changed back into its gaseous state or, possibly, made into methanol.

In the meantime, the battle over the natural gas pipeline bill and the manner in which Congress had been pressured to pass it had demonstrated in the most conclusive fashion the power of the Big Oil–banking interlocks that dominate American society. The people of the nation had had only a relative handful of responsible legislators willing to fight for them, and this dedicated band had been overwhelmed by the locust swarm of lobbyists flashing the power of big business and big money.

Bill Moyers, the onetime aide of Lyndon Johnson and in recent years one of the nation's most astute televsion commentators, summed it all up. "The two-party system is not only up for grabs—it's up for sale."

How Many Billions?

16

Official figures only hint at the way Big Oil has ripped off and savaged the American economy. The figures are impressive—they run into billions upon billions of dollars. Yet they only hint at the magnitude of the total scam. The heating oil rip-off that I have described—something never touched or investigated beyond the $3.4 billion gouge in thirteen months found by Representative Rosenthal's staff—would by itself total more than the entire swindle represented by official figures. Hundreds of reseller cases were never fully investigated. Others were dropped after halfhearted attempts to look. Still others were settled— a favorite practice of the Carter administration—by "penalizing" the offending companies and then giving them back a hefty portion of their "penalty" so that they could explore for more oil. It was as if a bank thief had been caught with $100,000 of his loot—and then, without being sent to jail, he was given back $50,000 to invest for his own profit and told to go and sin no more.

For what the official figures are worth—and they are not worth much—let's take a look at Representative Dingell's opening statement at a March 1981 hearing of his energy subcommittee. Dingell pointed out that his committee had held numerous hearings for the last five years on the enforcement conduct of D.O.E. In the last two years, he said, enforcement effectiveness "has improved to the point where the Special Counsel has alleged overcharges by the major oil companies of over $10 billion. Additional hundreds of millions of dollars in overcharges have been levied against the nonmajors."

Dingell was especially motivated in calling the hearing. President Reagan had just decontrolled all petroleum prices —an action that seemed to make old violations of controls almost moot. Furthermore, the president's controversial budget had slashed the heart and guts out of the halfhearted enforcement program Carter's D.O.E. had undertaken. David A. Stockman, Reagan's guru as director of the Office of Management and the Budget, had proposed to cut D.O.E.'s enforcement appropriation from $70 million to $13 million. On top of this, the administration had set a twelve-month deadline for the completion of pending audits and lawsuits against the oil companies. Since Big Oil, with its incredibly bursting multibillion-dollar treasury, could hire the most expensive legal counsel, it seemed inevitable that it could stall the government by one legal delaying tactic after another until the twelve-month deadline had expired.

"What this amounts to is amnesty for the oil industry under the guise of budget-cutting," declared Paul L. Bloom, D.O.E.'s special counsel in charge of enforcement. He left his office and returned to the practice of law. As he departed, he raised Dingell by $1 billion. He said his office had identified $11 billion in overcharges by thirty-four of the largest oil companies.

Avrom Landesman, who succeeded Bloom as acting special counsel, in a report prepared for Congress, said that $10.2 billion in alleged overcharges were still pending. The budget slash, he said, "doesn't provide resources to remedy" any more than about $3 billion of these charges. He added that "new enforcement actions won't be initiated even when substantial violations have been discovered." Landesman also pointed out that D.O.E. had never brought any charges stemming from its investigation of possible violations during the 1979 gasoline "crisis." Though D.O.E. officials confirmed that Landesman had made this assessment of the impact of Reagan budget cuts, the administration killed his analysis, deleting all of his critical comments from the budget document sent to Congress.

This was the background for the Dingell hearing, a last, diehard effort to demonstrate to the American people just how badly they had been rooked by the oil companies during the mid-1970s. Carter's Big Oil–manipulated D.O.E. certainly had been no crusading force, but under pressure by Dingell's committee and other congressmen, it had at least gone through the motions of taking some kind of enforcement action. Puny as the effort had been, it stood in marked contrast to the Reagan administration's blanket sanction of Big Oil in all its misdeeds.

Dingell harked back to the huge Daisy Chain scandal first exposed by Joe McNeff. This involved the hyping of "old" controlled oil to the highest going OPEC price through phony transactions in which the same oil was sold and resold many times through a chain of resellers, with the price written up at each stage along the way. Citing Energy Department documents, Dingell demonstrated that this particular scam went on at such a furious pace that the oil transfers far exceeded the capacity of the available pipelines to handle them. He made it clear that major oil companies, owners of the pipelines, were deeply involved.

Referring to D.O.E. data, he cited "the number of in-line transfers on two large pipelines—the Shell Rancho system and the Shipshoal system, which I believe is owned by Exxon."

As can be seen from the data turnover rate, a measure of the times each barrel of oil was resold has climbed dramatically since price controls were established in May 1973. The June 1979 data indicate that on the Rancho system the average barrel made available for in-line transfers was resold nearly twenty-one times. In all, the daily volume transferred was roughly six times the capacity of the pipeline.

Though none of the majors has been prosecuted, one criminal referral had been prepared for the Department of Justice involving Mobil—the second largest oil company in the nation—and American Petrofina and Tesoro. It alleged that in the single month of September 1978, these companies, in collaboration with a chain of resellers, wrote up the price of 8.3 million barrels of west Texas crude so that it came out of the end of the pipeline transformed into higher-priced "foreign" oil. The approximate overcharge for this one month was $43 million, but the D.O.E. referral pointed out that the reselling and marketing activities "were continuous for the period January 1978 through November 1978."

Among the participants in this reselling and mark-up scheme were two Carbonit firms—Carbonit of Houston and Carbonit International, headquartered in the Netherlands. The ties between Mobil and the Carbonit firms seemed especially close. Former employees of Mobil had become employees of Carbonit, and the president of Carbonit of Houston was a former vice president of Mobil. When D.O.E. attempted to get a deposition from him, he invoked his Fifth Amendment rights against possible self-incrimination to avoid testifying.

The D.O.E. referral to Justice had accused Mobil of engaging in one series of "apparently sham transactions" from which it made $4,272,315. The last paragraph of the D.O.E. referral to Justice read, "D.O.E. recommends the convening of a grand jury to conduct a thorough criminal investigation of the conduct described above." Yet no action had been taken by the Justice Department that had been so eager to harry Joe McNeff.

Representative Gore asked Jerome Weiner, director of special investigations, office of the special counsel for D.O.E., why nothing had been done. Weiner, who had himself sent the criminal referral to Justice, replied lamely that he had withdrawn the referral because "I concluded we needed further investigation with regard to the resellers and further investigation with regard to the majors."

Gore, the tiger of the Dingell committee, the man who had called the Daisy Chain scandal "probably the greatest criminal conspiracy in American history," wasn't happy at the way Mobil had escaped prosecution. He asked Weiner to explain why he had withdrawn his own recommendation. Weiner squirmed. He wouldn't let the name of Mobil pass his lips. What was happening? Gore wanted to know. Had more depositions been taken? Had bank accounts been examined? Or "is all that work left to be done and likely to be dropped?" Weiner protested that he could not respond in relation to "any specific matter" under investigation, not even by answering Gore's queries with a simple yes or no.

After some further questioning that failed to elicit any answers from Weiner and other D.O.E. witnesses, Gore, obviously frustrated, said angrily

My constituents during the time gasoline prices were continuing to jump upward were mad. A lot of them are still mad. If they knew that criminal violations took place on the part of major oil companies which resulted in money being illegally taken from them and cases were being dropped, they would not like it.

Doubtless, the congressman was right about the feelings of his constituents in Tennessee, but down in the heart of Texas it was another story. Dingell's committee learned of a conference held in Houston in early February 1981 by oil magnates and their legal mouthpieces. The lawyers advised their clients to "stonewall" the federal government, the committee heard. D.O.E. witnesses admitted that they knew of the meeting, but insisted they had no knowledge of what was discussed. Apparently, however, prior to the Houston conference, D.O.E. had had what was described as "two handshake agreements with large refiners to settle pending cases." After President Reagan decontrolled all petroleum prices and wielded his budget ax, the industry handshakers washed their hands of the agreement, assured of immunity.

The result was even worse under Reagan than it had been under Carter. Carter's industry-dominated D.O.E. had at least made some growling noises and enforcement gestures; under Reagan, whatever larceny may have been committed was legalized.

My files are loaded with clippings containing accounts of D.O.E. announcements of multiple-billion-dollar scams. CITE 7 OIL FIRMS FOR 1.7B OVERCHARGES, reads a headline from May 3, 1979. The story quoted Paul Bloom as alleging these overcharges in the last five and a half years: Texaco, $888.3 million; Gulf, $578 million; Standard Oil of California, $101.6 million; Atlantic Richfield, $42 million; Marathon Oil, $29.1 million; Standard Oil of Indiana, $24.1 million; Standard Oil of Ohio, $17 million.

A later headline from Sept. 19, 1979, reads MOBIL AND 2 OTHERS CONSENT TO REFUNDS. The Associated Press story explained that Mobil, while denying any guilt, had consented to refund some $13.8 million and to pay a $50,000 penalty. Two lesser firms had agreed to refund $8 million for overcharges on refined products, and in Houston a federal judge had fined Westmoreland Oil Development Corporation more than $1 million after the firm pleaded no contest to charges of overpricing.

The charges and the figures kept piling up. On November 1, 1979, Cities Service Company agreed either to make cash payments or forego price increases totaling $220 million and to spend another $150 million for increased domestic exploration. On November 15, Bloom accused Amerada Hess of overcharges totaling $99.1 million; Sun Oil, $26.6 million; Marathon Oil, another $17.4 million; Gulf Oil, another $1.7 million. He said these overcharges resulted from his audit of the books of the fifteen largest oil companies for the period 1975–76 only. On December 11 another headline read, U.S. ACCUSES GULF OIL OF MORE VIOLATIONS. The article in *The*

Wall Street Journal explained that D.O.E. was charging Gulf with an additional $486 million in violations and Standard Oil of California with another $64 million. The article noted that Gulf had settled previous charges of $578 million for $11.1 million and Standard of California had settled $101.6 million in alleged overcharges for $12.2 million.

This is the merest sampling of the flood of overcharging allegations that poured out of Bloom's D.O.E. office in the last months of 1979. The naive reader might have gotten the impression that Uncle Sam was really cracking down on Big Oil miscreants. Only close examination revealed that the D.O.E. watchdog's growl was far worse than his bite. For one thing, the charges resulted from alleged misdeeds of earlier years: D.O.E. had not even looked at 1979—the year of the phony gasoline "crisis," the most profitable year in Big Oil's history. For another, the settlements engineered by Paul Bloom and his assistants amounted to little more than gentle taps on the wrists. No Big Oil moguls were prosecuted for criminal actions—their companies were "penalized" for amounts that represented only fractions of the alleged over-charges. For instance, the $11.1 million Gulf settlement represented only 2 percent of the alleged overcharges. They were allowed to walk away protesting their innocence and saying they had settled only because they wanted to avoid further annoyance. It was much like letting a godfather of the Mafia off with a severe lecture and a fine that he didn't even feel because his illegal profits were so enormous.

In the oil company cases, however, there was added another even more offensive obeisance to the power of Big Oil. In case after case, a large part of the "penalty" wasn't a penalty at all. The companies were permitted to take the money and reinvest it in their own businesses. The Carter administration's rationalization for such leniency was that we were in such desperate straits that the oil companies' promise to put some millions of their ill-gotten gains back into their own businesses was really the nation's gain.

A few examples will show the way this "tough" settlement policy worked. On Nov. 9, 1979, the Energy Department and Phillips Petroleum Company signed a $200-million consent agreement. Major components of the agreement were: (1) cash refunds of $3 million to nine of Phillips's crude oil customers, (2) purchases on the international market of $22 milion worth of crude oil to be turned into heating oil, with prices frozen at the existing level, (3) a *loss* of $67.5 million in *future* gasoline price increases and (4) a commitment to spend $100 million for new and accelerated oil and gas exploration. In other words, out of the $200-million overcharging "penalty," Phillips was allowed to keep half for its own future profit on its promise to do more with it.

This was the pattern in case after case. The huge headlines about multibillion-dollar overcharges were followed months later by mouselike announcements of fairly innocuous settlements. The most outrageous example of this procedure was the handling of the Standard Oil of Indiana (Amoco) case. I have in my files a letter Paul Bloom sent to Sen. Jacob K. Javits after Javits had inquired on behalf of one of his constituents about what was happening. After describing the Cities Service and Phillips agreements, Bloom wrote

Finally, Standard Oil of Indiana (Amoco) agreed to a consent order totalling $690 million in remedies. A $100 million refund was established. Twenty-nine million dollars was to be returned to specified large volume purchasers of middle distillate products and $71 million will be placed in an escrow account to be distributed under procedures being developed by the Department. The remaining $410 million of the Amoco settlement amount will be directed to new and expanded projects undertaken by the company.

In other words, Amoco got back for its own use and profit nearly two-thirds of the headlined $690 million "penalty" assessed by the government for its overcharges. Some penalty!

The Justice Department made some angry noises about Bloom's handling of these agreements. Associate Attorney General John H. Shenefield objected that the Energy Department was exceeding its authority by agreeing to drop lawsuits against the companies as one of the conditions for the settlements. He argued that only the Justice Department had the right to take such actions. Energy Department lawyers replied that Justice had been slow to do anything and that it was just jealous of Bloom's high-profile enforcement drive. The oil companies, laughing up their collective sleeves, taunted that Justice couldn't do anything about it because the Secretary of D.O.E., a cabinet officer of stature equal to the attorney general's, had signed the agreements.

After Bloom left office and Reagan appointees took over, there was a definite slackening of even Bloom's moderate enforcement activity, partly as the result of the drastic budget cuts that had eliminated the jobs of auditors necessary to put together the cases. In July 1981, the Reagan D.O.E. obtained an $82.5-million settlement from Chevron for overpricing and improper accounting since 1973. But this was in settlement of charges amounting to $417 million. Cases against Atlantic Richfield and Union Oil were still pending, D.O.E. said.

Representative Gore was decidedly unhappy. At a meeting of the House energy and commerce subcommittee, he charged that D.O.E. was making "pious but empty promises" about prosecuting all violations of the old oil price controls.

A draft General Accounting Office report of late February 1982 substantiated Gore's charges. G.A.O. in a March 1979 study had found D.O.E.'s Energy Regulatory Administration had had "very little success in resolving the alleged violations." In its new report, G.A.O. found no significant improvement. In fact, some conditions were worse. E.R.A., the report said, "has experienced decreasing company cooperation and as one consequence, D.O.E. had ninety-two subpoenas outstanding as of December 1981, fifty-seven of which were over six months old." E.R.A.'s "staff morale has continued to suffer" as experienced auditors left or were "removed from responsible management positions. . . ." And, finally, "E.R.A. has had little success in negotiating settlements of alleged violations with companies."

The record made it clear that Big Oil could get away with almost anything and that there was no incentive in the political hierarchy in Washington to stop it. There had been little enough during Carter's tenure, but there was virtually none under Reagan, whose philosophy, molded during those years he was shilling for General Electric, is that bigger is better.

Indicative of the virtual paralysis of government where Big Oil is concerned was the abject abandonment of the Federal Trade Commission's antitrust suit against eight major oil companies. Started in 1973, the suit had charged "collusive action" by Atlantic Richfield, Exxon, Gulf, Mobil, Shell, Texaco, Standard Oil of Ohio and Standard Oil of Indiana. The original F.T.C. complaint made charges similar to those brought by the attorneys general of the western states in the Pad V suit. It had accused the companies of maintaining and reinforcing monopoly power over oil refining in large parts of the United States since 1950, of limiting the supply of crude oil to independent refineries and of following a system of price posting that kept oil prices at artificial levels.

The F.T.C. had started out with the hope of breaking up the vertically integrated oil companies—that is, the Big Oil corporations that handle everything from the wellhead through refining to distribution to local dealers and gas stations. It had hoped to split the companies into separate production, refining, pipeline and distribution units. But after nearly eight years of legal battles with batteries of high-priced oil company attorneys—and faced with an increasingly hostile attitude in Congress and the incoming Reagan administration—the F.T.C. abandoned the struggle in February 1981. The antitrust case went into limbo and was finally killed in September because, the F.T.C. said, it could continue in the courts indefinitely before a decision could be reached.

During the years of struggle with Big Oil attorneys, the F.T.C. had obtained massive amounts of documents from oil company files. Documents obtained from Mobil alone are said to fill more than one hundred large cartons. Obtaining access to such internal documents and memoranda, however, could hardly be called a victory, for the F.T.C. had been forced to make an agreement that hamstrung it or anyone else from making any use of the material.

Representative Dingell's committee sought the F.T.C. files after the antitrust case was dropped. It obtained them, but the court agreement under which the files were obtained provides that before any use can be made of the "proprietary" information in them, the oil companies must be notified so that they can go into federal court to obtain an injunction against disclosure or use of the information.

Such is the power of Big Oil. A committee of Congress has access to its secrets but remains bound and helpless. It may peep, but its lips are sealed—and so are the files.

17

The Sin of

Synfuels

Exxon has a vision for us. Take the whole northwestern corner of Colorado and turn it into a "national energy zone" in which "normal rules" would not apply. Tear apart the landscape to mine the shale; crush the shale and heat it in a retort to get out kerogen, a low-grade form of crude oil. And so develop a synthetic fuels industry capable of producing 15 million barrels a day by the start of the next century.

There will naturally be problems. Water, for one. It will take vast amounts of water. Even Exxon admits that water is a precious resource in the shale mesas of Colorado. The upper limit that could be processed with available water would be 1.5 million barrels a day. This doesn't stop Exxon. Take the water from the Missouri and Mississippi river basins. Expensive but not impossible, says Exxon.

Of course, water won't be the only problem. The cooking of the shale would spew out huge amounts of air pollutants —hence the reason "normal rules" won't apply in Exxon's national energy zone. And there will be an enormous waste disposal problem. Since shale will expand by about one-fifth in the heating process, a mound of slag larger than the miles-long, miles-wide crater from which it came will be left behind for disposal. In some twenty years—the estimated life of the Exxon project—some 400 million tons of spent shale will be produced. Salts and toxic materials contained in the rock besides kerogen will be leached out by rain- and snowfall.

Such leachate contains flourides, boron, arsenic, molybdenum, selenium and alkaline salts—all posing a threat to the Colorado River system. Even after refilling the huge craters caused by mining, this mountain of leftover leaching materials will pose enormous disposal and environmental problems.

This vision of an energy-independent future—all in the national interest, of course—was unfolded to the editors of the *Washington Post* at a luncheon meeting with Exxon Chairman C.C. Garvin, Jr., in 1980. Exxon, which gave no indication it was worried very much by environmental problems, had been circulating a draft analysis of its view of the nation's energy future, and Garvin in his luncheon meeting with the *Post's* editors had expanded upon the plan for turning all of northwestern Colorado into an energy zone in which Exxon's project alone would produce 8 million barrels a day. Other, similar projects would produce the remainder of Garvin's envisioned 15-million-barrels-a-day output.

Disclosure of Exxon's blueprint for the future "scared the bejesus out of everybody," one Colorado energy expert said. Environmental and business groups, county and state officials were galvanized into protest by Exxon's plan for their future. Many recognized that shale oil development might be coming, but virtually no one was reconciled to the mammoth development pictured by Exxon or to the waiving of environmental rules to accommodate it. "If I'm still in the Senate, they could do that only over my dead body," Colorado Senator Gary Hart said after being informed of Garvin's remarks.

Senator Hart is not dead, but it must be assumed that he is recumbent. For Exxon is not just talking—it is acting. And its plan for a massive shale-oil processing plant along Parachute Creek was approved by the Reagan administration and implemented by the U.S. Synthetic Fuels Corporation, a federally funded government agency.

The synthetic fuels corporation was a concept of

President Carter. It was created to fund and encourage the development of synthetic fuels to relieve our dependence on foreign sources and to provide energy by the year 2000 when, according to Carter and Big Oil, the nation's larder of fossil fuels would be barren. The independent corporation was to be funded over the years by $88 billion, from which it would make seed-money grants to projects it deemed the most promising.

During Carter's administration, before the synthetic fuels corporation could be organized and get into action, D.O.E. itself had funded various projects. It had experimented with wind turbines to produce electricity. Three were erected, on the mainland, in Puerto Rico and in Hawaii. The most successful, Makani Huila, erected on Oahu's north shore, generated 1,170,980 kilowatt hours in seventeen months through November 1981, saving 2,000 barrels of fuel oil.

Another pilot project funded by D.O.E. tested the theory of ocean thermal energy conversion (OTEC). This process uses the temperature difference between warm surface water and colder deep water to generate electricity. A state of Hawaii research vessel, moored off Ke-ahole Point on the island of Hawaii in 1979 and tested throughout 1980, proved the validity of the theory, but further development work needs to be done.

D.O.E. also funded research into solar energy in collaboration with some major electric utilities. The use of the photovoltaic cell to generate power from the sun had furnished energy for our space probes, and with D.O.E.'s aid, it was tested out on a larger scale that indicated it could generate significant amounts of electricity for utilities, reducing their reliance on nonrenewable resources like oil, gas and coal.

These innovative approaches died when the Reagan administration took office. Reagan immediately announced his intention of abolishing the entire Department of Energy, a goal he hasn't yet achieved, and he appointed an oil man, Edward E. Noble, to head the U.S. Synthetic Fuels Corporation, with the avowed intention of dismantling this as soon as possible.

In a speech before the National Petroleum Refiners Association in Hilton Head, S.C., on September 14, 1981, Noble announced his intention of cutting off the entire synthetic fuels program at the end of a $20-billion first phase. He declared, "There is a provision for a potential Phase II authorization of $68 billion—but, gentlemen, I can tell you that I'm going to do my darn'dest to see this job is done without this monstrous second phase."

With an oil man in charge of synthetic fuels development, with an administration in the White House that is wedded to the biggest of big business, it became virtually inevitable that the only projects to receive government aid would be those sponsored by such financially needy organizations as Exxon and Union Oil. The administration had no interest in the development of renewable sources of energy. It cut off solar funding and aid for such projects as wind turbines and OTEC. One highly knowledgeable congressional aide said

Reagan is even opposed to further hydroelectric development. He has no interest in anything except nuclear power. The lobbying that goes on by the big firms is incredible. If you think small companies should play a role and have a chance to be financed, if you think renewable fuels should be given a chance—after all, the sun is always there and solar energy is renewable; coal is not—it is all very disturbing. The Reagan administration seems afraid of having anything to do with 'the environmental crazies,' as they call them. The indications are they will deal only with the big corporations.

This was a perceptive comment. The first three projects approved by D.O.E. and Reagan himself were the multi-billion-dollar kind that benefit Big Oil. Two were shale oil projects in Colorado—one the Exxon-Tosco Colony Oil

Shale Project at Parachute Creek, the other, a similar Union Oil project. The third grant was for a huge coal gasification plant in Beulah, N. Dak., sponsored by a consortium of major companies headed by American Natural Resources, of Detroit.

The Exxon-Tosco project, especially in view of Garvin's prescription for the future of Colorado, aroused the greatest controversy. The so-called Colony Project was conceived in 1974 when Atlantic Richfield and little-known Tosco—the name is an acronym for The Oil Shale Company—planned to build a 10,000-barrel-a-day plant on the mesa above the middle fork of Parachute Creek about fifteen miles west of Rifle, Colo. At the time, however, oil prices were controlled, there was no federal financial backing, and environmental hazards were still a government concern. As a result, the project was left in limbo.

Then, on August 1, 1980, Exxon came along. It bought Atlantic Richfield's 60-percent share of the project for $400 million. In the fall of 1980, about the same time Garvin was explaining his grandiose vision to the *Washington Post,* Exxon and Tosco began to build a road from the creek up past the planned mine opening and on to the top of the mesa more than 1,000 feet above the spot where the processing plant was to be located.

With Exxon's bulging billions kept discreetly in the background, Tosco applied to the Synthetic Fuels Corporation for aid in developing the project. All Tosco sought was a loan guarantee for $1.125 billion to help it finance its 40-percent share of the $3.154 billion project. The amount it sought was 75 percent of its $1.493 billion obligation, the maximum percentage the government was allowed to grant under the law.

Budget Director Stockman opposed this beneficence. So did Sen. William Proxmire, a caustic critic of government extravagance, and Sen. Harrison (Jack) Schmitt, a New Mexico Republican. Both Proxmire and Schmitt were on the

Senate Banking Committee and had held hearings on the Exxon-Tosco proposal.

The two senators introduced a resolution trying to kill the loan guarantee to Tosco. In the debate, Schmitt pointed out that this was the first synthetic fuels project to be approved and "as such it is setting a significant precedent that other such companies will expect to be able to follow." He emphasized that in

the last fiscal year federal loan and loan guarantee programs preempted an even greater share of domestic credit resources than did the deficit. . . . Because of their unobtrusive nature, loans and loan guarantees have become an easy vehicle through which to siphon off vast amounts of capital to favored lobby and interest groups, and unlike the recent exercise in budget cutting that affected many of lower or moderate income groups, many of the Federal lending programs are directed to individuals and enterprises that are not economically disadvantaged and who have access to private sources of credit.

Then he asked a question.

Why if the promises of the project are so overwhelming does this venture deserve $1.12 billion in guarantees from the government? . . . No information has been provided regarding the question of Exxon's willingness to further finance it, or Tosco's efforts to find other financing. Why should they want to if the government can underwrite dream loans for them, anyway?

Proxmire lashed out at the Exxon connection and said the first question was whether the taxpayers should be called on to "assist a project largely supported by one of America's largest and most profitable corporations." He added

A D.O.E. representative testified last week before the Banking Committee that D.O.E. never asked how much aid was really necessary for the project to be undertaken nor did the agency ask if Exxon would fund the entire project if no loan guarantees were provided for Tosco.

The Exxon-Tosco program called for the production of kerogen crude oil, which would have to be further refined to make it usable, at a cost of $34 a barrel, the OPEC bench-mark. But many were skeptical. Many thought there would be heavy cost overruns on the huge project, and they doubted that shale oil could be produced at a reasonable price. Others looked at Tosco's balance sheet and felt that the federal government might be left holding a $1.125-billion bag. Tosco's report for the first two quarters of 1981 as the issue was being debated showed that the company's total assets were only slightly more than $1 billion, matching the size of its loan, and its operations had resulted in a loss of $1.22 a share.

What would happen in case of default and the collapse of the project? The Colorado Open Space Council, the Natural Resources Defense Council, Friends of the Earth, and the Environmental Policy Center pointed out in a state-ment filed with the Senate Banking Committee that only the government would lose. Under the agreement, Exxon and Tosco would be required to relinquish only the "licenses, patents, technologies and proprietary rights which are neces-sary for the completion and operation of the project. . . . The government will receive no royalties or additional access to the technology. Tosco retains all of its patents, licenses and proprietary rights."

The groups concluded, "It appears as if D.O.E. negoti-ators did a much more effective job of protecting the assets of Tosco than the capital of the federal government."

The partiality of D.O.E. to Big Oil interests is hardly surprising to anyone who has followed the story this far; and so it comes as no shock that what Big Oil wanted Big Oil got. Despite the efforts of Senators Schmitt and Proxmire, the $1.125-billion sweetheart deal with Tosco was approved.

The award to Tosco was topped by a second grant—this a $2-billion loan guarantee to the American Natural Resources Company for its proposed Beulah, N. Dak., coal gasification plant. The project had had a checkered history. President Carter in 1980 had announced with great fanfare a $250-million conditional guarantee to get the project started, but even this support hadn't been enough to rally the necessary private financing. The plant was estimated to cost at least $2.5 billion and was designed to produce 125 million cubic feet a day of synthetic gas made from lignite coal. Potential buyers of the gas complained about is inevitably high cost, and doubts about the product's marketability in view of huge natural gas discoveries seemed to doom the proposition.

The Reagan administration at first looked dubiously at the so-called Great Plains proposal. Some administration officials contended that the plant was designed to produce the wrong product at the wrong time. They argued that, with the current oversupply of natural gas, the synthetic gas produced by the plant would not be marketable. Instead of producing synthetic gas, they argued, the plant should focus on liquid fuels.

Then lobbyists went to work and whipped up prosynfuel sentiment among both parties in Congress. Administration attitudes began to change. In July 1981, Energy Secretary James Edwards said that the coal gasification project "has to be considered a marginal one from the standpoint of competitiveness," but at the same time he said he was strongly urging the President to approve the loan because of "the need to demonstrate the technology." What Edwards was urging, however—and what D.O.E. and Reagan ultimately approved—was not a pilot plant to "demonstrate the technology" but a full-scale commercial venture.*

* An indication that critics were right and the Reagan administration wrong came in March 1982. Panhandle Eastern Corporation announced it had scrapped plans for building a huge gasification plant in Wyoming. The plant would have been the second largest in the United States, exceeded only by the Great Plains project. Panhandle Eastern cited as reasons for abandoning its project soaring construction costs, high interest rates and the world oil glut "that casts additional doubt on the marketability of the product." (See *The Wall Street Journal,* March 25, 1982.)

The third project in this first synfuels series involved a Union Oil shale oil development in the same Colorado region as the Exxon-Tosco plant. Union did not get a loan guarantee but something that was perhaps more valuable—a contract for the government to purchase its product and a price guarantee. Under the contract, the government was obligated to spend up to $400 million in price supports to Union to enable it to go ahead with its project.

The big-business campaign to tap the federal till for additional billions of dollars continued. D.O.E. through 1981 had been funding five large-scale synthetic projects. They were called demonstration projects, but actually they were so huge they were commercial ventures. The five projects had received a total of $870.3 million—and all were still in the design stage when the Reagan administration took office. Finding that the industrial sponsors were unwilling to pay more than a small percentage of the total cost (in three cases the sponsors had contributed only 15 percent or less), the administration decided to stop this drain on the federal treasury.

Four of the projects promptly died, but the fifth—a monstrous solvent-refined coal process known as SRC I— survived and became the focus of a battle in Congress for continued funding. Sponsors of the plant, to be built in Newman, Ky., were the Wheelabrator CleanFuel Corporation, a subsidiary of Wheelabrator-Frye, and Air Products & Chemicals Corporation. Plans called for SRC I to consume 6,000 tons of coal a day and to produce the equivalent of 20,000 barrels of oil in the form of a solid boiler fuel, anode grade coke and a hydrogenated liquid boiler fuel. The plans also called for the sponsors, through their operating corporation, the International Coal Refining Company, to put up only $90 million while the government supplied $1.772 billion.

The project, which had the backing of the state of Kentucky and the two senators from the state, had grown like Topsy. Its cost had been estimated originally at $685 million; by 1979, this had ballooned to $2.289 billion, and by 1981 to $4.574 billion. To make the project seem financially feasible, the estimates of returns from the sales of the plant's three products had escalated right along with the costs. Estimates of revenues from the first five years of operation had been raised by the sponsors from $801 million to $2.802 billion—a staggeringly optimistic forecast 250 percent higher than the original figure.

The General Accounting Office and the National Academy of Sciences looked at this picture and were not enchanted by what they saw. G.A.O. reported that liquefaction developers had told it there was a much greater risk in moving from "small-scale pilot plants to a demonstration effort, especially with commercial-scaled coal reactors. . . ." The National Academy of Sciences found that data for the design of a commercial plant could be obtained from a unit using only 250 to 600 tons of coal a day. G.A.O. added that even Wheelabrator had informed it that a 2,000-ton-a-day plant "would be the largest size for which commercially available equipment can be practically employed. . . ."

Yet the campaign in Congress mounted to commit the federal treasury to the sponsorship of a plant three times that size. Some senators and congressmen balked.

Representative Moffett, chairman of the House subcommittee having oversight of the Synthetic Fuels Corporation, held a hearing at which Representative H. Joel Deckard of Indiana objected that "prevailing winds from western Kentucky carry over my state." He expressed concern about the "toxic wastes and pollutants"—substances like arsenic, cadmium and mercury—that might be produced by the plant. "The synfuels crude produced from the liquefaction process is carcinogenic," he declared.

Subsequent House and Senate debates focused, however, more on economic feasibility than on environmental concerns. Representative Ottinger contended that "we should use our coal supplies for liquid fuels" and raised the ghost of methanol. "We can make methanol at the present time for 40 cents a gallon commercially," he said, citing testimony before the subcommittee on science and technology. "It comes to about $15 to $16 per barrel in manufacturing costs, and we can use it to propel our vehicles or for stationary purposes."

Experts had concluded, Ottinger said, that the 40-cent-a-gallon methanol could be delivered to drivers at the pumps for 80 cents. "Why it is not being done is a great, great mystery," he said. "Methanol producers will not produce it in quantity because they say there is no emergency. . . . The automobile companies say they will not produce this modified oil product."

(The automobile companies, of course, do not produce fuel. What Ottinger obviously referred to was that they had resisted any motor modifications to permit use of alternate fuels. Only Ford has designed and tested a methanol-powered engine.)

Is it possible that Big Oil could have given Ottinger the answer to his "great, great mystery"?

After the House had voted to appropriate another $135 million (on top of the $100 million the government had already contributed) just to cover further design costs, the issue went to the Senate. Senator J. Bennett Johnston, Jr., led the proplant forces. He argued that "not even Exxon" would support a project of this size and that we have "hundreds of years worth of coal, a resource base so vast it dwarfs Saudi Arabian crude oil."

The proposed plant, he said, would produce a clean product that could be used in boilers "so that you have no sulphur emissions, so that you have virtually no ash, so that you have no problems with acid rain, which is now threatening many of the lakes of the Adirondacks and other areas. . . . Some areas have water so acid that it can not be used for drinking purposes."

Canadians, Johnston added, "have proved up their process and are now producing liquid fuels at a price that approaches seventeen dollars a barrel."

Sen. Jennings Randolph, representing a great coal-mining state, supported the Kentucky project and harked back to the Synthetic Liquid Fuels Act signed by President Franklin D. Roosevelt in 1944. This had authorized the secretary of the interior, acting through the U.S. Bureau of Mines, to develop new technologies to produce methanol, ethanol and other liquid fuels from coal, shale oil and agricultural products.

In 1947, he said, an Interior Department memorandum foresaw that a federally subsidized synthetic fuels industry would be able to produce two million barrels of oil a day within the next five years. In 1949, the Bureau of Mines managed two synthetic plants in Missouri "which turned coal into an assortment of petroleum products, including gasoline for vehicles and diesel fuel for railroad engines. The cost, . . . said the government experts, was 3 to 4 cents higher per gallon than the same product from crude oil. . . .

Arguing that alternative sources of energy must be developed, Randolph, like Ottinger, raised the spectre of methanol, saying, "Methanol as an automobile fuel should be encouraged. Promotion of this fuel source would . . . significantly reduce the demand for conventional petroleum."

Opponents concentrated on the enormous financial commitment being sought from the federal government and doubted the revenue estimates of the sponsors. Sen. Don Nickles ridiculed the idea that the plant, if completed, would be able to run for five years at 100 percent of capacity as the sponsors optimistically predicted. Even so, he calculated that the costs of the plant and its operation during this period would be so high that "if you divide by the number of barrels of total production, we are talking about a cost of $76 per barrel. . . ."

I seriously question whether or not this product . . . will be able to sell at $76 a barrel to break even . . . and if we are going to start paying back the taxpayers for their investment . . . it is going to have to be more valuable than $76 a barrel.

Proxmire objected to what he considered the looting of the public till to benefit private interests. In a gentlemanly clash with Sen. Walter Huddleston, a strong supporter of the Newman project, he said, "What bothers this Senator very much is that here we have a situation where Uncle Sugar—the taxpayer—puts up everything. How much does this private contractor put up? Very, very little until 1985."

HUDDLESTON: The sum of $90 million.
PROXMIRE: Very minimal, while the federal government puts all the money in and takes all the risks.

In the end, the Senate voted fifty-seven to forty against a Proxmire amendment that would have killed off the project. So the government was committed to another $135-million handout, though it remained uncertain whether the Newman SRC I project would ever be completed.

The whole debate raised two important issues, the second of which no one seems to want to think about, much less to mention. The first issue concerns the government's role in promoting synthetic fuels. Many think, like Professor Robert Williams of Princeton University, an energy expert, that the government's legitimate role has been distorted by the issuance of multibillion-dollar grants. He thinks government funding should be confined to research and development, not to commercialization.

Some alternative technologies are very promising. If you do basic and applied research in areas like gasification from biomass instead of coal, you can get much further with the expenditure of far less money. I think all of the alternatives—solar, the photovoltaic cell, methanol—should be pursued, but I'm very skeptical about government deciding in advance which is going to be the best in the long run. Government should be supporting research, not commercialization.

The synthetic fuels program as it is presently being managed, however, is supporting commercialization at public expense and favoring the huge Big Oil and financial interests that increasingly have the nation by the throat. It is a situation that suggests the second and, in the climate of the times, the unthinkable alternative: If the government and its taxpayers are to bear the major part of the multi-billion-dollar costs of commercializing new technologies, then shouldn't they reap a proportionate share of the profits? Failing that, then shouldn't the government run the projects itself? This, of course, raises for the Establishment the horrible prospect of government competing with private industry in the energy field, but the choice seems to lie between this kind of TVA-like competition and the oligopolistic, monopolistic domination of the American economy that we have now.

About this domination there can be no question, but it was never more obvious than in the continuing maneuvers of Big Oil once the first huge synthetic oil projects had been underwritten by the federal government. Tax break after tax break was slipped through Congress to benefit the Big Oil shale oilers. First, in addition to billion-dollar aid from the government, the companies get the standard 10 percent depreciation writeoff on their own investments. Second, they engineered another 10 percent tax credit for "pre-retorting" activities—that is, for mining before oil production is begun. Third, they slipped through Congress in the eleventh hour of the 1981 session another 10 percent tax credit for "post-retorting" activities—in other words, the cost of the cleanup. It all added up to a 30 percent tax give-away off the top.

Even all this did not satisfy Big Oil. In the spring of 1982, still salivating over the prospect of synthetic oil fortunes, it attempted another steal. It tried to slip through

Congress a bill that would give several major oil companies 40 percent of all the federally owned shale oil lands in the West. Lobbyists for the oil industry—Gulf, Exxon, Amoco, and Standard Oil of Ohio—worked feverishly to secure passage of this colossal give-away in a bill sponsored by Sen. John W. Warner. A similar measure had snaked quietly through the House. Passage of Warner's bill would ratify the deed. In calling for swift action, Warner beat the national defense drums, saying: "Our overall national defense posture is no stronger than our energy posture."

Behind this flag-waving was the grasping hand of insatiable greed. Warner's bill would have leased mineral lands to the oil companies at 50 cents an acre, the price that was being paid in 1920. Companies leasing comparable private lands are paying hundreds of dollars an acre. Warner's bill set no level for royalty returns to the federal government. Some earlier approved leases have guaranteed only a 2 percent royalty compared, for example, to the 25 percent Conoco is paying private ranchers on shale oil leases. Warner's bill would have left it up to Secretary of Interior James G. Watt to determine what royalty should be paid— and Watt hasn't been known to step on a big business toe yet. The bill set no time limit for performance; once Big Oil acquired the lands, it could hold them until it was bloody well certain the time was ripe for huge profits. This is totally unlike provisions in leases on privately owned lands, where, if the leasing company doesn't perform within a specified period of time, the rancher, wanting his potential 25 percent royalty, can sue to void the lease.

As in the case of the Alaska gas pipeline bill, there was a smothering silence about all of this until the *Washington Post* made a belated and partial discovery in late March of 1982. Warner's bill had been scheduled to slide through the Senate before the Easter recess, but the *Post*'s disclosure and some hastily aroused opposition resulted in a postponement of the vote, a delay that may have been fatal.

"The oil companies are coming in and literally trying to steal this land, but nobody seems to be paying any attention," Senator Metzenbaum complained. "They are worse than the robber barons, and the Republicans are doing their bidding."

But the robber barons were about to get an unexpected comeuppance, and this put a different complexion on Senator Warner's shale oil giveaway. "The bill looks dead in view of what has happened," one of Metzenbaum's aides said, "but don't count too heavily on it. The bill is still on the Senate calendar, and there is no question the industry definitely wants it bad. There is always the possibility that it could be slipped through in the eleventh hour of the closing session by a device like attaching it as a rider to some other measure."

What had happened to give the "robber barons" and Senator Warner this momentary check was the sudden collapse of the first billion-dollar-plus shale oil boondoggle: the Exxon-Tosco Colony Shale Oil Project.

The skepticism of opponents about the fiscal sanity of this venture begun to be borne out almost as soon as the $1.12 billion guarantee to Tosco had been made. As early as March, 1982, anxious tremors ran through the federal bureaucracy when Exxon jumped its cost estimate for the project from $3.124 billion to $5 billion. Government officials began to worry about the security of the Tosco loan and about whether the shale oil to be produced would be price-competitive with more conventional fuels. The government threatened to revoke the loan guarantee to Tosco; but, after some protracted negotiations, it decided to stand by its commitment.

Then, on May 2, Exxon pulled out the rug. Overnight, it announced that it had decided to abandon the entire Colony project. Exxon said its reappraisal had convinced it the project would cost "more than twice as much" as the $3.1 billion originally estimated, and it could not be certain that even this ballooned figure represented the end of the overruns. "Economics no longer support" the investment, said Randall Meyer, president of Exxon's U.S. unit.

Exxon had come to this conclusion as the result of a second appraisal, this one dealing with crude oil prices and worldwide energy use for the rest of the decade. This study indicated that prices would not rise as much as had been expected, and analysts estimated that crude oil prices would have to climb to $50 a barrel for the plant to make money.

The collapse of the Colony project sent shudders through the synthetic oil industry. "Synthetics have been indefinitely postponed, maybe never to get off the ground," said John H. Linctblau, president of the Petroleum Industry Research Foundation.

Tosco, which disagreed with Exxon's reasons for abandoning the project, nevertheless came out of the mess a big winner. Under its contract with Exxon, its big brother was obligated to pay it $380 million for its interest in Colony should the project be abandoned. Thus, Tosco could pay back the federal government for the loans it had so far received and come away with an estimated $100 million profit.

This Exxon-Tosco debacle did not affect Union Oil, the other active shale oil entrepreneur in the Parachute, Colo., area. Union Oil, it will be recalled, had been satisfied with a $400-million government guarantee to purchase its shale oil product. If Union's oil should ultimately cost $50 a barrel, the government is obligated to take it—and pay for it at the expense of the American taxpayers.

Exxon's scuttling of the Colony project seemed to doom for the time being, at least, Chairman Garvin's fantasy about turning all of northwestern Colorado into an "energy zone." And it seemed to threaten the boom town of Parachute with the spectre of becoming another Western ghost town.

Before Exxon descended on the mesa, Parachute had been a crossroads town with 360 residents. In two years, it had grown to 1,200. Some 2,100 workers had been drawn to the new energy-job mecca, most of them living in a mobile-home village that Exxon had created for temporary living quarters while it set about building an entire new subdivision that it called Battlement Mesa.

Battlement Mesa was to be the incarnation of Exxon's dream. It was to be virtually a small city, housing some 25,000 residents on a 3,300-acre site. Some 8,000 buildings were to rise on the mesa: condominiums, $70,000–$80,000 private homes, 10 churches, 8 schools, shopping centers, a $5.5 million recreation center, an 18-hole golf course.

Two schools, one being built by Exxon, the other by Union Oil, were nearing completion. One shopping mall had been built and was about to open. Some of the 432 apartments under construction had been completed and were renting for between $450 and $600 a month. Some of the new homes had been sold to workers who had plumped down their life's savings and assumed heavy mortgages, believing they would have well-paying jobs and security for years to come.

Then, on Sunday evening, May 2, radio and television carried the catastrophic announcement: Exxon was pulling out. Everything stopped.

The next day was marked by turmoil, ugly violence. Furious workers hurled stones at buildings. Some drove their trucks through fences, smashing them. Others staged a mini-run on banks to get their savings out. Still others descended on liquor stores and became ingloriously drunk. The sheriff called up extra deputies and closed down one liquor store in an effort to restore order.

"It came as a great shock," Mayor Floyd McDaniel, owner of the largest food store in Parachute, told me that Monday morning. "The community is at the mercy of the oil companies about what they are going to do. They come in and we have to gear up to take care of things, lay sewage lines, etc., and then when they pull out, we have to take it on the chin. That has been the history of synthetic fuels, up and down, up and down.

"Union Oil is going ahead with its project because it has that price guarantee, and that will help some. But a lot of people are going to be terribly hurt. I'm afraid some who paid $70,000 to $80,000 for new homes with mortgages on them are going to lose everything. The owner of the one shopping mall that is almost completed will probably hang on. But a developer who was going to build a second mall—land clearing had just started—will probably try to pull out of his lease. Construction had begun on a couple of light industrial parks. That will cease. The two schools that are being built are nearly finished and will be completed. But instead of having a town of some 20,000 to 25,000 people, we'll probably have only one-third or perhaps only a quarter of that. There is some talk about Mobil becoming active in the area, but nobody knows what Mobil plans to do. I just don't know what the future holds."

It remains to be seen whether the collapse of the Colony Shale Oil project represents more than just a temporary setback. Though the first reaction in financial circles was that such projects were doomed, Exxon's chairman, C.C. Garvin, Jr., was not ready to abandon his "national energy zone" dream. At the one-hundredth anniversary annual meeting of Exxon stockholders in New York on May 13, 1982, he said the decision to stop work had been "part of an across-the-board reappraisal of major projects made necessary by the unsettled state of the energy markets."

Exxon, he said, would take an after-tax writeoff of $125 million, but he made it clear the company was not writing off the entire $920 million it had invested in the project.

"We put it in mothballs," he said. "We'll continue to do the necessary research work. It was not a question of shutting it down and forgetting it."

In other words, if there should be another energy crunch, if oil prices should be driven up to $50 a barrel, or perhaps

$76 a barrel, Exxon would be prepared to resume work on the Parachute mesa and pursue its "national energy zone" dream. What was clear at the end was what had seemed obvious from the start: synthetic oil from shale can be marketable only if crude oil prices are driven into the stratosphere at incalcuable cost to our industrial economy.

18

An Alternative We Don't Want

Two high columns stood on the Washington Mall near the Capitol in the last days of April 1979, just as the nation was suffering in the throes of the Big Oil–manipulated gasoline shortage. Linking the two continuous fermentation columns was a series of four tanks. In a distilling process that took an hour's running time, this demonstration plant turned six bushels of corn into fifteen gallons of alcohol, and it yielded in addition enough methane gas to fuel the distilling process.

The still had been designed by Dr. Paul Middaugh, of the University of South Dakota, and it was intended to demonstrate—just at a time when gasoline pumps were running dry, prices were soaring and motorists were furious —that America had the resources and capability to run its cars on fuels other than gasoline. The demonstration was staged by a coalition of public interest groups, aides of Ralph Nader and South Dakota farmers connected with the American Agriculture Movement.

For three days the still ran on the Washington Mall, producing 2.5 gallons of alcohol for every bushel of corn used. At the end of the distilling process, it also yielded a high-protein mash valuable as cattle feed. Senators and congressmen witnessed the three-day demonstration, and many came away impressed, wondering, Was the use of ethyl alcohol a way to break the back of OPEC and give us an indigenous, renewable motor fuel?

The prospect was certainly tempting. Eliot Janeway, the economist, was advocating "drowning OPEC in a sea of alcohol." In the past, two of America's most famous inventors, Alexander Graham Bell and Henry Ford, had advocated the use of alcohol as a motor fuel. Bell had called it an extremely clean burning, efficient source of energy, and Henry Ford had been so impressed that he had designed Model T carburetors to run on either gasoline or ethanol.

Ethanol differs from its alcohol cousin methanol in that it is produced, not from coal, but from all kinds of agricultural products, cannery wastes and biomass—from substances like sugar cane, corn stalks, wood chips and even walnut shells. Ethanol is less corrosive than methanol, it mixes better with gasoline, and it is a valuable booster to upgrade unleaded gasoline into higher-octane premium unleaded. Ten percent of it mixed into a gallon of gasoline produces a product called *gasohol*. In 1979–1980, with the gasoline panic sweeping the nation as the result of the supposed Iranian "shortfall," there was a lot of public excitement about gasohol.

Three years later, circumstances had changed radically. The world oil glut, the recession and dramatically reduced consumption (American imports by March 1982 were down to 3 million barrels a day from their high of over 8 million in 1977) had dissipated the air of crisis and so, to a large extent, the American public's interest in renewable fuels. With regular gasoline dropping down to almost a dollar a gallon, why should one worry?*

Many experts believe, however, that this euphoric mood is the delusion of the moment—that it will pass and we will

* The American Petroleum Institute reported that imports for the week ending February 26, 1982, were down to 2.7 million barrels a day, the lowest level in seven years. The previous week imports had been only 3.5 million barrels a day, compared to 5.1 million barrels for the same week in 1981. Conservation, switching to alternative fuels and the recession were cited as causes for the lower imports. (See *The Wall Street Journal*, March 4, 1982.)

be brought back to earth with a hard jolt. This will certainly happen if D.O.E.'s projections of $2.20 gasoline (or $5.50 counting for inflation) turn out to be accurate. The situation is filled with uncertainties. No one can be certain how long the present oil glut will last. No one can be sure that there won't be another war in the Middle East (this seems at times dangerously likely), in which event vital oil supplies could be cut off again. And so there are those that insist the nation should not turn its back on ethanol fuel.

Big Oil, of course, has other ideas, as its fascination with gigantic shale oil and coal gasification and liquefaction plants shows. These ventures can be commercially viable only at that $76-a-barrel plateau (or more) that Senator Nickles had ridiculed in the debate over the Newman, Ky., SRC I proposal. On the other hand, gasohol, or pure alcohol fuels, though not price competitive in a market of abundance and falling prices, could look positively cheap if $76-a-barrel oil and $2.20 gasoline lie in our future along the road Big Oil is paving for us. It is an uncertain scenario at best, and one that suggests the nation might give sober thought to technologies that offer fewer drawbacks than the ripping apart of the landscape in the West for shale oil and coal.

My attention was first attracted to the alcohol alternative in 1979 by Scott Sklar, the Washington area director of the National Council for Appropriate Technology. Sklar had been an assistant to Senator Javits specializing in energy matters. When I first spoke to him, he had been running all around Washington for eight months, driving a 1964 Rambler Classic four-door sedan fueled on pure ethanol. The old Rambler, he said, perked along fine and gave him twenty miles to the gallon.

Sklar described his head-butting experiences with the Energy and Research Development Administration of Carter's D.O.E. during the ten years he had served as Javits's assistant.

"Every time I suggested something, they had the answers why it wouldn't work," Sklar said.

They were the experts, and I figured they must be right. Some Nebraska gasohol advocates gave me some statistics that seemed to show that gasohol did work, that it gave even better mileage than pure gasoline, but when I went back to ERDA, they shot those figures down, too.

Then a friend of mine in the Ford Motor Company told me about experiments Volkswagen had run. I found Volkswagen had conducted literally hundreds of tests using different blends of alcohol and gasoline, and even pure alcohol. Their tests showed that alcohol really worked well. I went back to ERDA, and they gave me the old story about BTUs—they never seemed to be able to think about anything but BTUs—and they showed me some very inadequate data that had been furnished by the Petroleum Institute on tests that had been run on just seven cars. This made me mad.

Sklar did some further research, and he found that, though alcohol has only about two-thirds the BTUs per gallon that gasoline does, it has "a much greater thermal efficiency. It is 1.4 times more efficient than gasoline in its direct conversion to power." Sklar was angry enough to decide to try alcohol fuels for himself. He found it necessary to make only a few minor adjustments to the carburetor of his old Rambler and to add a small container of gasoline, connected to a button on the dashboard, that would squirt a jet of gas into the carburetor to start the motor on cold mornings. Ethanol, he found, will not start readily when it is cold, but once the spurt of gas got the motor started, the engine would warm up quickly and run the rest of the day without trouble. Sklar made quite a game out of riding congressmen and senators around Washington to demonstrate to them how well his alcohol-fueled car worked.

The Mall demonstration and Sklar's scooting around town in his alcohol-fueled Rambler led to a series of hearings by Representative Ottinger's science and technology sub-committee. Ottinger opened the hearings by saying he and other congressmen had seen the Mall demonstration and had been impressed.

"One problem I have with the government," he told Department of Agriculture witnesses who were critical of ethanol, "is that it is always looking for something that will produce huge profits."

He referred to a Midwestern farmer who was using a solar still to make alcohol out of his surplus corn and running his farm equipment on it. "If you add a little savings here and a little there, it all adds up," Ottinger pointed out.

It was a philosophy that had been anathema to both D.O.E. and the Department of Agriculture. They had argued that it cost more energy to produce ethanol than ethanol could deliver; that food production would be jeopardized if agricultural products were used for fuel instead of food; that ethanol would rust and damage motors. It came as a great surprise, therefore, when D.O.E. did a sudden aboutface and weighed in with a strong endorsement of alcohol fuels. Alvin L. Alm, assistant secretary in charge of policy and evaluation, declared that use of alcohol could extend our energy resources and reduce gasoline consumption.

"We have found that alcohol fuels can be produced with a positive gain as long as natural gas and petroleum are not used in the distilling process," he said. "We believe that further economies could be affected in technology."

Alcohol was being produced by conventional methods at the time for between $1.20 and $1.50 a gallon, but Alm was convinced it could be produced at substantially lower prices. He estimated that if the nation made full use of its agricultural, waste and garbage resources it might be possible by the 1990s to "extend the petroleum supply of the country by 80 percent, maybe more."

Alm's testimony surprised Rep. Dan Glickman, who said it seemed to represent a complete turnaround from what both D.O.E. and the Agriculture Department had been saying. He thought the national interest required a united effort by Agriculture, Energy, Treasury and Defense. As soon became clear, however, the national interest was being subordinated to the private profit interest. The point emerged in the testimony of two Gulf Oil officials.

Gulf, it should be said by way of preface, had exhibited unusual initiative for a petroleum company. For seven years, at a research cost of approximately $9 million, it had conducted experiments to develop the very alcohol-conversion technology in which Ottinger's committee was interested.

The effort had been centered in Kansas City, Missouri, under the direction of Dr. George H. Emert. Dr. Emert had worked with a petrochemical plant, not one designed for fermentation and, therefore, far from maximally efficient. He had tested municipal solid wastes (garbage), agricultural wastes and pulp and paper wastes. The last two he had found ideal because they contain a lot of cellulose. His petrochemical plant, processing a ton of solid wastes a day, made forty gallons of alcohol, but Dr. Emert's laboratory experiments indicated that an efficiently designed plant should yield seventy to ninety gallons from the same tonnage.

Dr. Emert and his staff of thirty researchers had mapped out for Gulf an ambitious schedule. It called for the construction of a fifty-ton-a-day demonstration plant to be brought on stream sometime in 1980 or early 1981. Following that, Dr. Emert's schedule called for the construction of a full-scale commercial plant that would cost $112 million to build and would produce 50 million gallons of alcohol a year.

The witnesses for Gulf at this point had the task of explaining why the great oil company, having spent $9 million and brought the project to the threshold of success, had decided to abandon it. The witnesses were George E. Huff, vice president for science and technology, and William P. Moyles, vice president for administration and development.

Huff testified about Emert's research and said Gulf would be able to bring a commercial plant into operation in 1983. Even allowing for four more years of inflation, Gulf estimated such a plant could produce alcohol selling for $1.45 a gallon, with a 15-percent profit margin after taxes.

Here Moyles took over. He said Gulf had made "a commercial decision" not to go ahead. Gulf, he said, had committed $2.5 billion to exploration for new natural gas and petroleum sources, and "we're trying to live with our cash flow." Thus, he indicated, the $112 million needed for Dr. Emert's commercial plant on top of the budgeted $2.5 billion would be the straw that broke the camel's back. "It was just a question of where we were going to put our money," he said.

Representative Gore boiled at this.

"Is Gulf going ahead on its own?" he asked sharply.

"*No, sir!*" Moyles declared emphatically.

Huff interjected that he thought the project feasible, that it could be done, that it should be done—but by someone else to whom Gulf would be willing to sell its technology.

"I find this fascinating," Gore said. "You have done all the research, and you have discovered, apparently to your horror, that it works."

In a continuing exchange with Gulf officials, Gore marveled that one wing of Gulf was enthusiastic about its alcohol project—and another wing wanted to sell it off. "I wonder if it might have occurred to you that the production of alcohol threatens your oil holdings?" he asked.

Moyles replied that it might be ten or fifteen years before the alcohol plant contributed to company earnings—this despite Huff's earlier testimony that such a plant would yield a 15-percent after-tax profit if it were completed in 1983.

Gore commented that "we're just plain crazy" and that "we're out of our minds" if we rely on the oil companies to develop alternative sources of energy. "This seems to be a classic example of what is wrong," he said.

Huff protested that Gulf *was* interested in developing other sources of energy. He said the company had committed $500 million to develop oil from shale and coal (but the $112 million for Dr. Emert's 15-percent-profit fermentation plant would be just too much). Gore hopped on this alibi.

"But the big difference is that you can control the reserves without buying the material from farmers," he said. He pointed out that the oil companies owned 50 percent of the nation's coal reserves, and when Huff protested that the oil companies didn't own the shale lands, the federal government did, Gore fired back

Yes, and the federal government owns all the offshore oil lands, too, but it's you and your Six Sisters who are developing them. What about the national interest? I say again I think we're crazy if we leave the nation's fuel future in the hands of the big oil companies.

It was one of the more illuminating exchanges at congressional hearings that I attended, and afterward I had several long talks with Dr. Emert. Gulf's withdrawal had wrecked his timetable, and it was clear that his envisioned full-scale commercial plant wasn't going to be turning out 50 million gallons of alcohol a year by 1983. But Dr. Emert remained a fully committed man. "We need all the alternative sources we can find and we need them now," he said.

He drew this picture at the time (1979): In the previous year, we had consumed 128 billion gallons of gasoline; we made only 600 million gallons of alcohol for beverage and other purposes. To make enough alcohol to fulfill gasohol's 10-percent requirement, we would have to turn out some 12 billion gallons.

This was not an impossible task, as Dr. Emert saw it. He emphasized that alcohol boosts octane ratings on regular unleaded gasoline and yields an average of 5 percent more mileage, reducing by this amount the number of gallons needed for driving. As for raw materials, Dr. Emert calculated that between 8 and 10 billion gallons of alcohol could be produced from solid wastes alone—from the billions of tons of garbage accumulated by our towns and cities, from wastes that now create landfill problems and leach into our lakes and streams, causing widespread pollution. This, however, would be only a beginning.

The nation would still have excess corn and wheat, especially if acres kept out of production for price support purposes were cultivated. This would give the farmers of the Middle West a new market, a new source of revenue. Other agricultural changes could easily be made. More crops with higher yields for alcohol purposes than corn could be grown —crops like sugar beets, sugar cane and sweet sorghum (perhaps the best source of all). In addition, there were the untapped resources of some 400 million acres of woodland.

Dr. Emert placed a high value on this last resource. He recognized the difficulty of relying on grains alone because crops fluctuate with the weather and prices with the market. Corn could sell at times for $2 a bushel—less than it cost the farmer to produce it—at other times for $5 a bushel, at which price it became uneconomical for use in alcohol plants. The solution, as Dr. Emert saw it, was to design fermentation plants to take advantage of whatever supplies were most available. The additional cost of a multiple-resource plant was relatively insignificant, Dr. Emert said, and this was just the kind of plant that he had designed for Gulf—one that could use garbage or grains, cannery wastes or wood byproducts.

The use of ethanol as either a fuel enhancer or an un-blended motor fuel naturally presents some problems. One arises from the fact that alcohol is such a clean fuel it flushes out the muck in a gas tank and lines, clogging filters. Unless filters are changed frequently, the car will sputter and stall, but if proper precautions are taken, this can be avoided.

Ethanol has been tested extensively in Brazil. Since Brazil has to depend on petroleum imports, and since it grows vast amounts of sugar cane ideal for making ethanol, it began a program in 1975 to encourage the use of the alcohol fuel with the goal of having at least 45 percent of Brazil's automobiles running on pure alcohol by 1985. It now appears that this won't happen.

Bugs in the system have disillusioned many Brazilian drivers. There was the problem of starting up cars in cold weather (a real one unless that squirt of gasoline is used as a primer). In addition to this, some drivers found that long, continued use of ethanol ate away at engine parts so that costly repairs had to be made more frequently. Both Volkswagen and Ford engineers in the country have been developing metal-coating methods to protect engine parts that come into direct contact with ethanol. Auto engineers insist that most of the complaints come from owners of older-model cars that were poorly converted before some of these difficulties were corrected. They point out that some drivers are delighted with the performance of alcohol-fueled cars. One cab driver exulted to *The Wall Street Journal* about the performance of his 1981 Volkswagen Beetle. "It performs better than a gas engine," he said. "I can even go up hills in high gear."

In the United States, there has been no attempt to pursue the possibilities of ethanol on such a national scale. Indeed, the enthusiasm for the fuel, originally generated by farmers in the Midwest, quickly cooled for two reasons. First, alcohol fuels bumped headfirst into the $88-billion Carter synfuels program; second, the sudden oil glut of 1981–1982 with the resultant drop in gasoline prices removed, for the moment at least, the incentive to experiment with alternative fuels.

The synfuels program, which might have been expected to promote alternatives like alcohol, became their almost instant enemy. By its very nature, ethanol production does not lend itself to huge, concentrated refineries; it is of necessity a more dispersed enterprise, with fermentation plants located close to various sources of raw materials. Hence, it could not be controlled as easily by Big Oil—and the synfuels program from the start was in the grasping hands of Big Oil through its minions in D.O.E.

Even as the first $20-billion phase of Carter's synfuels act was being put in place in the late summer of 1980, Barry Commoner, the environmentalist and renewable fuels advocate, denounced it at a hearing of the National Alcohol Fuels Commission. D.O.E. had just made its initial $250-million commitment to the Great Plains gasification project, and Commoner was outraged. Gasification from coal, he warned, is "a very nasty process," producing vast amounts of toxic materials. He added, "The synthetic fuels legislation . . . is a very bad piece of legislation and could well become a national disaster."

He illustrated by citing the ecological havoc wrought in a downstate county of Illinois of which he had personal knowledge. This area, he said, had possessed some of the blackest, richest farm soil in the county. Then strip miners came in, tore up the landscape for coal and moved on, leaving ruin behind them. This was the fate he foresaw for much of the West as the federal government seemed hell-bent on committing itself to only the most grandiose multibillion-dollar shale oil and coal gasification projects.

It was a perceptive analysis. The $112 million needed to bring Dr. Emert's ethanol plant into commercial production by 1983 seemed like a penny-bank investment compared to the $1-billion to $2-billion amounts that the federal government was soon bestowing on projects sponsored by Big Oil.

When the Reagan administration took office—even more committed to Big Oil than Carter, if that is possible—one of the first things it did was to try to wipe out the ethanol program. D.O.E. had given preliminary approval for 706 million dollars' worth of loan guarantees for eleven planned alcohol fuel projects expected to produce 365 million gallons a year, more than twice the current alcohol production for motor fuel. When the Reagan budget-cutters took their knives to the program, Congress balked and ordered the funds restored. Since then, one of the projects has been abandoned, but the others are being carried forward.

From this, it can be seen that all circumstances have combined to put a decided damper on ethanol production as a supplementary energy stock, but curiously enough, all of these circumstances combined have not been enough to kill off the whole program. James D. Stearns, director of the Office of Alcohol Fuels in D.O.E., acknowledged that the competitive picture has become more unfavorable as gasoline prices have declined dramatically. But corn prices have dropped also, and Stearns feels that the ethanol alternative is making slow but steady progress.

He emphasized that, in 1981, 71 million gallons of ethanol were used as an octane booster in 710 million gallons of gasoline. Ethanol was especially popular in three states— California, Iowa and Michigan. "Two plants in California are producing ethanol from cannery wastes," Stearns said. "Even walnut shells and almond shells can be turned into ethanol, and something that was pure waste can be turned into dollars."

There is another and huge source of waste that can also be turned into dollars, Stearns added. The government in its farm price-support program, annually buys and stores hundreds of millions of bushels of grain; millions of bushels are left over from one year to the next. "Grain in storage deteriorates," Stearns explained.

After a time, weevils get into it, and it spoils, until it gets to the point where it is no good at all for food purposes. But this wasted grain would make a perfectly good feedstock for an ethanol plant, so it is not just a question of growing more corn to produce ethanol, but of using what otherwise would be wasted.

This problem was compounded by the record-breaking 1981 grain crop. The nation's farmers produced more than 8 billion bushels of corn and nearly 3 billion bushels of wheat, a harvest so bountiful that grain elevators couldn't store it all. So the problem that confronts us is not that of diverting grain from food to produce ethanol, but of making efficient use of resources that otherwise would be wasted.*

Californians, according to Stearns, have been especially enthusiastic users of ethanol-boosted fuel. "The state is the largest importer of foreign cars, especially Japanese makes," he explained, "and buyers of the high-compression Datsun 250, for instance, have found that super lead-free gas with ethanol as its booster is a much superior fuel. It lets them zip around in their sporty cars at a much swifter pace."

Texaco, as Stearns emphasizes, is the one major oil company that, from the beginning, embraced gasohol and marketed it at its pumps. Though gasohol sales have fallen off in the East and Texaco has stopped selling it at some of its stations, at the end of 1981 the company was still marketing gasohol at 1,600 locations in nineteen states and the District of Columbia. In addition, in collaboration with the Corn Bureau Cooperative, Texaco had built an ethanol plant in Pekin, Ill. Test production began at the end of 1981, and when the plant swings into full operation, it is expected to produce 50 million gallons of ethanol a year.

Stearns pointed out that half a dozen farmers' cooperatives, with six hundred to nine hundred members each, are going ahead with plans to build their own ethanol plants

* The 1981 harvest was so huge that the Reagan administration is asking wheat farmers to reduce the acres they plant by 15 percent and corn farmers to cut by 10 percent if they are to qualify for price-support benefits. The 1981 harvest yielded 2.8 billion bushels of wheat and 8.2 billion bushels of corn. As of February 1982, there was still a surplus of 1.06 billion bushels of wheat and 2.02 billion bushels of corn. With both domestic and foreign sales lagging, with storage facilities inadequate, with early prospects pointing to another bumper crop of winter wheat, the administration was forcing farmers to cut back on the acreage they plant in 1982—or to plow some crops under—in order to avoid an unmanageable glut. (See *The New York Times,* April 7, 1982.)

without "asking for any handouts from government." The farmers feel, he said, that the 4-cent-a-gallon federal tax break given ethanol is sufficient inducement. "They think," Stearns explained, "that they can assure themselves of feedstock supplies and that they can produce and market their own fuel."

Margaret Livingston, of the Martin-Haley Company, Washington consultants, is the adviser to the farm cooperatives. The movement started in 1979 in North Dakota, she explained, and the plan calls for six cooperatives to build their own ethanol plants and for a central marketing cooperative to handle distribution. Construction of the first plant in Hankinson, N. Dak., is scheduled to begin in August 1982, with the plant coming on line in late 1984 or the first of 1985. It is expected to produce 50 million gallons of ethanol a year.

"The farm cooperative idea is very good," Ms. Livingston says. "The farmers are very involved. They are in business to get an extra market for themselves. They will use corn as a primary feedstock, and the plants will produce ethanol to be sold to refineries for blending into premium-unleaded gasoline. The only byproduct will be high-protein silage to be used for animal feed."

Such cooperative programs and other, on-farm fermentation plants can help American farmers, who consumed 3.5 billion gallons of oil in 1978, reach for energy independence. Diesel engines have operated successfully on blends of 30 percent ethanol and 70 percent diesel oil if lubricating additives are used and minor engine modifications made. Tests indicate that a 50-percent blend is possible. Stearns pointed out that the Germans "are making considerable progress in the use of both methanol and ethanol in diesel engines. This kind of use in tractors on farms and in farm machinery could consume a significant amount of alcohol." And save a considerable amount of oil.

Despite the 1981–82 oil glut, there is a widespread belief that this superabundance of petroleum is a temporary phenomenon, that there will be tighter markets in the future —and higher prices, as Big Oil's subsidiary, D.O.E., has predicted. OPEC nations, meeting in March 1982, demonstrated a willingness to cut back on production in order to maintain prices. Though this initial reduction was relatively small (a cut of 750,000 barrels a day), the united action seemed like a possible forerunner of more drastic moves in the future if they became necessary to dry up the international market and maintain high prices. Over all hangs the ever-present threat of another Middle East war.

Thus the possible future need for alternative energy sources cannot be disregarded. Alternative energy, of course, takes many forms. One that is coming into widespread use throughout the nation is the siphoning of methane gas from old landfills. Getty Oil, of Los Angeles, pioneered this process as long ago as 1975 and now operates three of nine such plants in the nation. The methane drawn from landfills is used primarily for electrical generation. A lot of this landfill gas has been recovered, but a lot more lies buried in the nation's dumps. D.O.E. has estimated that, just by tapping the two hundred largest landfills in the country, 55 billion cubic feet of methane could be produced each year.

Fuel to drive America's cars and trucks, however, seems most likely to have to come from other alternative sources— methanol and ethanol. Ethanol presents fewer technical problems, and Dr. Emert, who perfected the Gulf Oil process, thinks the nation is still going to need large amounts of ethanol. He scoffed at the government's obsession with huge shale oil projects in the West. "My studies show," he said, "that shale oil selling prices will have to be about forty times what it costs to produce ethanol."

Dr. Emert has continued his work at the University of Arkansas to which Gulf transferred its rights on a royalty-sharing basis. The university research center holds four patents covering various stages of the ethanol process, and Dr. Emert has become a director of a firm called United Bio-Fuel Industries, Incorporated, of Richmond, Va. Bio-Fuel is building a plant in Petersburg, Va., on a seventy-three-acre site along the Appomattox River that includes a fifty-three-acre city landfill. The process perfected by Dr. Emert calls for the use of several enzymes to break down cellulose and convert it to fermentable sugars. Yeast is then added to transform the sugars into ethanol. The ethanol is extracted by a patented process that permits reuse of some of the enzymes and yeast for further fermentation, thus reducing costs. The plant is programmed to produce 50 million gallons of ethanol a year at a cost varying from 70 cents to $1.15 a gallon, depending on financing and the kind and cost of resource materials. Gulf Oil, under its contract with the University of Arkansas, will receive 22.5 percent of whatever royalties the university collects for the use of its patents.

Commenting on his work and the outlook for alcohol fuels, Dr. Emert echoed the views of Professor Williams of Princeton, regarding the role of the federal government. He too thinks the government should concentrate on promoting research and development programs, not on funding commercial enterprises.

The greatest impact on the whole program has been Reagan's new approach to the budget. Many projects are being diminished because of the lack of funds. Many laboratories have closed. Others have had to cut back their R & D work. Fortunately, we've been able to continue, though we've had to cut back on some programs, too.

This federal impact has been tremendously significant. There are men who have spent their lives working on these projects, and they were on the threshold of bringing things to fruition when suddenly the threshold was pulled out from under them. It was a terrible mistake, and I can't imagine anyone in Reagan's position being so shortsighted. Of course, I suppose there may be political reasons that I am not aware of, but I don't know how long it will take for us to recover from the damage that has been done.

Or how long it will take for the political process to recognize the folly of following the Big Oil road to the ultimate dead end.

Whoops!

19

The Pacific Northwest was in seething revolt in the winter of 1982. Protestors marched, picketed, threatened runs on banks. They wore large buttons proclaiming, "I'm Irate" and "Resign or Recall"—a demand aimed at Washington state's Public Utility Commissioners. "These people are very mad, and they are almost uncontrollable," said Senator Susan B. Gould, chairman of the energy and utility committee of the Washington state senate.

The fury was aimed at the Washington Public Power Supply System, derisively known as "Whoops." It was caused by Whoops's attempt to build five huge nuclear power plants —and the resulting overruns that had left the agency $7.7 billion in debt with none of the systems completed.

Washington Public Power electric bills had doubled—in some cases tripled—and, as the costs of the nuclear albatross mounted, they threatened to double or triple again. Customers in white-hot rage were refusing to pay their bills, they withdrew some $600,000 in deposits from Washington's largest bank, a backer of the utility system—they even threatened a run on the bank.

The revolt was sparked, not by long-haired youths or fanatic environmentalists, but by some of the most conservative citizenry. Indeed, the nucleus of one irate ratepayers association had been compiled from a mailing list maintained by the National Rifle Association.

The chaos in the usually stable Northwest was just the most dramatic illustration of a national verity: Nuclear energy had finally been discredited. Not because it posed environmental hazards, not because the possibility of a nuclear disaster worse than Three-Mile Island might wipe out an entire countryside—but simply because it had turned into a financial debacle.

Almost from the moment Big Oil and OPEC had conspired to produce the petroleum shortages and the price escalations of the 1970s, a succession of national administrations had hailed nuclear power as the one sure way to solve some of the worst of our problems. The nuclear genie, in this view, could bring us into an all-electrical age, supplying cheap and inexhaustible power. Despite the evidence of the times, this dream of the nuclear miracle still possessed Ronald Reagan.

But with the public and business communities alike, nuclear had lost its credibility. No nuclear plants had been ordered since 1978, and many that had been ordered, even some on which construction had been started, were being abandoned and closed down. The fact was that nuclear power had turned into a multibillion-dollar nightmare—a nightmare for consumers whose tripling bills were costing them more for electricity than they had ever imagined possible, a nightmare for utilities that found themselves burdened with skyrocketing construction costs and plants that, even when completed, were frequently breaking down, a nightmare for stockholders and bondholders who saw their investments threatened. What was happening in the Pacific Northwest demonstrated the manifold problems that were leaving the nuclear giant moribund.

Whoops had started construction of the five nuclear plants that had been expected to take care of the growing needs of the Pacific Northwest well into the next century. Eighty-eight municipal utilities had become participants in the project through "take or pay" contracts to buy the power —a so-called hell-or-high-water deal that obligated the utilities to pay the costs even if the plants were never completed, even if they never got any power from them.

The construction costs escalated; delays were encountered; overruns mounted into billions upon billions of dollars. By late 1981, Plant No. 4 was only 25 percent completed and Plant No. 5 was only about 16 percent finished. Construction costs on the two plants had tripled. They had already eaten up $2.25 billion, and it was estimated it would cost $12 billion more to complete them. At this point work was stopped, but costs continued. It was estimated that another $343 million would be needed to pay off contractors and mothball the skeletons. It was estimated, too, that the eighty-eight tied-in utilities would have to pay off the already incurred $2.25 billion—and an additional $4.65 billion in interest over a period of years, all for plants that would never be built and energy they would never receive.

Even this was not the end of Whoops's problems. Plants one, two and three were still under construction. Only Plant No. 2 was nearing completion, and another $556 million would be needed to finish it. Plant No. 3, which had cost $1.3 billion so far, was only 44.8 percent finished, and it seemed extremely likely that work would have to be stopped on both No. 1 and No. 3 since preliminary surveys indicated the Northwest might not need as much power as had been thought when the plants were started.

It was a mess. Nuclear power was a mess almost everywhere. In California, more than a thousand Pacific Gas & Electric consumers marched on the governor's office in Sacramento. A restaurant in Petaluma, Calif., dramatized the protest against high electric rates by serving all of its meals by candlelight. Southern California Edison, one of the nation's largest utilities, in a dramatic about-face turned its back on nuclear power and announced it would rely in the future on such renewable sources as solar, geothermal and wind power. The Nuclear Regulatory Commission (N.R.C.)

estimated that as many as thirty-seven nuclear plants either planned or already under construction had been cancelled by utilities around the nation.

Even nuclear plants that had been finished and had been functioning for years were turning into economic liabilities. The disaster at Three-Mile Island that had spread a cloud of radioactive steam over the Pennsylvania countryside in 1979 was the most publicized near catastrophe, but it was not the only failure. Three years after Three-Mile Island, the billion-dollar-plus cleanup of its No. 2 reactor still goes on. In addition, the No. 1 reactor, which General Public Utilities, the holding company owning Three-Mile Island, had expected to get back into service remains deactivated because thousands of steam generator tubes used to carry nuclear-heated water had deteriorated so badly that they must be replaced—a task that it was estimated in February 1982 would take at least another six months.

In the meantime, the staggering costs of this less sensational nuclear debacle are being passed on to utility consumers in Pennsylvania and New Jersey. Three utilities —Jersey Central Power & Light (JCP&L), Metropolitan Edison and Pennsylvania Electric, partners in GPU's Three-Mile Island adventure—have been given hundreds of millions of dollars in rate increases for the costs they incurred and in rate adjustments for buying more expensive replacement energy. Jersey Central and Metropolitan, both teetering on the edge of bankruptcy, are seeking further bailouts from state public utility commissions—at the expense, of course, of consumers.

Jersey Central Power & Light's problems, however, did not end with Three-Mile Island. It had a second nuclear problem on its hands. Its Oyster Creek nuclear plant in Ocean County, N.J., after being in operation for eleven years was breaking down with increasing frequency. The plant had to be closed for seven months on one occasion in 1980; then, later in the same year, it had to be closed down twice more for additional repairs. The closings cost consumers $750,000 a day for replacement energy purchased from costly oil-fired plants. Even these repeated shutdowns for repairs didn't put Oyster Creek back in A-1 operating condition. In January 1982, the plant had to be closed down again for "repairs and modifications." It was expected to remain off-line for twenty of the next thirty-six months. The estimated cost to 704,000 residential, commercial and industrial consumers: an extra $15.63 to $19.50 per month for each month the plant is out of action. These were extra tariffs that would be piled on top of the hundreds of millions of dollars already being paid by consumers as a result of Three-Mile Island.

Radioactive leaks in nuclear power plants, resulting shutdowns and exorbitant charges to consumers for replacement energy were like a plague spreading across the nation. New York's huge Consolidated Edison, one of the nation's largest utilities, was having problems similar to those of JCP&L. Con Ed's Indian Point No. 2 reactor functioned on an average only eleven days a month during 1981—and the costs of repairing and maintaining the silent monster boosted the electric bills of New York consumers to 2.51 cents per kilowatt hour, more than twice the 1980 figure of 1.22 cents.

At this point, inevitable questions arise. What has happened? What went wrong with the vision of cheap and inexhaustible nuclear power that had seemed so beguiling? A close study of one case—not by any means unique, on the contrary, typical (that is the significance of it)—supplies some answers.

The illustrative experience involves Florida Power & Light, the state's largest utility. Beginning in the early 1970s, FP&L constructed a four-plant generating complex on Turkey Point abutting Biscayne Bay, some twenty-five miles south of Miami. Two of the plants, Turkey Point No. 3 and Turkey Point No. 4, were nuclear. Turkey Point No. 3 went into operation December 14, 1972; Turkey Point No. 4 came on line September 3, 1973. The reactors had cost about $120 million apiece to build and were supposed, according to the nuclear mythology of the day, to function for some forty years.

As early as 1974, only two years after the first plant went into operation, FP&L apparently began to have some problems. In one filing with the Atomic Safety and Licensing Board of the Nuclear Regulatory Commission, the utility admitted that it had made changes in the system. In 1974, it said "the sodium phosphate secondary water chemistry treatment for the steam generators was converted to an all-volatile chemistry treatment. Following the conversion, in 1975, certain corrosion-related problems such as denting of steam generator tubes began to occur." The problems became steadily worse, and by Dec. 13, 1977, they had become so bad that FP&L requested permission from the Nuclear Regulatory Commission to rebuild the two plants.

In its decision on the application, the licensing board summed up the Turkey Point difficulties.

The wastage and denting phenomenon have led to . . . several instances of coolant leakage through cracked tubes. As of November 1980, tube plugging for various reasons has resulted in removing about 20 percent of the steam generator tubes in Unit 3 and about 24 percent in Unit 4 from continuing service. . . .

Denting occurs when a buildup of rust pinches the tubing at its support plates inside the generator; wastage refers to the destruction of the tubes' inner walls by rust. These are the tubes that carry nuclear-heated water to the generators where steam is used to produce electricity. There are some 3,200 tubes in each generator, so defective ones can be sealed without much loss of power—up to a point. Plugging at the Turkey Point plants had been so extensive, however, that the generators were operating at only 65 percent of capacity.

The Turkey Point units had been designed by Westinghouse and had been equipped with machinery produced by the company. Some twenty nuclear reactors in use in the nation had also been designed by Westinghouse, and many had experienced similar tubing and cracking problems, as had some generators built by Combustion Engineering. Virginia Electric Power's Surry plant, also designed by Westinghouse, had encountered such serious problems that it had had to be rebuilt before FP&L made application to retrofit Turkey Point Nos. 3 and 4. Three other retrofitting applications were pending before the N.R.C. when the FP&L case came before it.

The public usually has no knowledge of plans that may affect lives—and certainly will affect pocketbooks—until the deed is done. The details are hidden behind a screen of technical jargon contained in correspondence between the utility and the N.R.C. Public hearings are considered a waste of time. Nuclear power plants are required to maintain a document room, usually in a nearby university, where all the arcane details are available to the public if anyone should take the trouble to wade through them. Usually, no one does.

In the FP&L case, everything was proceeding smoothly on the greased bureaucratic track without any public busy-bodies interfering. FP&L applied for permission to rebuild its two Turkeys; the N.R.C. made some noises; FP&L explained; FP&L was granted permission to go ahead. No public hearing was held.

At this point, one of the most unusual characters in the saga of nuclear misadventures derailed the smoothly functioning bureaucratic machinery. His name is Mark P. Oncavage. He is a sixth grade music teacher in Dade County, the French horn his specialty. He had, unfortunately for FP&L, an inquiring mind, and he was taking some night courses in environmental subjects at Florida International University. It just so happened that FP&L, no doubt to its eternal regret, had selected this particular university as the repository for those nuclear files that no one was expected to examine. The utility, of course, had never heard of Oncavage.

While researching a term paper, Oncavage came across the exchanges between FP&L and the N.R.C. He knew little about nuclear techniques, but he was curious. And so he began to wade through the intricate correspondence—and the more he waded, the less he liked what he saw.

FP&L's plans, as approved by the N.R.C., called for ripping apart the two nuclear plants and rebuilding them at an estimated cost of $425 million (double what they had cost originally and a figure far below some later estimates that put the ultimate rebuilding cost at $700 million). The six old, radioactive generators from the two plants were to be stored in a floorless concrete building only five feet above the level of Biscayne Bay. The utility also planned to dump 100,000 gallons of its primary coolant, the radiator fluid of the reactor, into its outdoor cooling canals leading to Biscayne Bay. Oncavage began to worry. What would happen if a hurricane whipped the Florida coast and sent waves crashing over those radioactive generators? Would the dumping of the coolant cause a biological backlash in the coast's food chain? And what was to be done with all the other radioactive materials that would be torn out of the dismantled plants?

As Oncavage later told me, "I saw a lot of problems that no one was paying any attention to. Not the power company, not the government, not the N.R.C." He decided to try to stick a big legal spike into the smoothly meshing wheels of FP&L and the N.R.C.

Oncavage and some friends formed an organization, Floridians United for Safe Energy (FUSE). He got the help of lawyers Neil Chonin, Bruce Rogow, Martin Hodder and Joel Lumer. The N.R.C. had not done an environmental impact study before approving FP&L's plans. Oncavage and his lawyers concluded that it should have. Though more than a year had passed—the deadline for intervening in such a case—Chonin filed an appeal for Oncavage with the Atomic Safety and Licensing Board, an arbitration panel that mediates N.R.C. disputes. The three-member board flew to Miami in 1979 to hear Oncavage's objections, much to the disgust of FP&L, which didn't like this horn-blower's horning his way into its apparently settled case.

"Back when the first nuclear plants were built," Oncavage later explained, "everybody considered nuclear power a terrific idea. They didn't understand the hazards of working in a plant that was gradually becoming completely contaminated. Now it's extremely dangerous to workers who have to go in and try to take it apart."

Chonin, representing Oncavage, and Lumer, representing FUSE, argued before the licensing board that FP&L's application to rebuild the Turkey Point units should be held up for at least a year until it could be determined whether the repairs already completed in the Surry plant in Virginia had actually corrected the flaws.

"There is no proof that Westinghouse has solved the problem," Lumer said. "Westinghouse thinks it has, but is that just guessing? There is no guarantee that the plants, once repaired, won't break down again."

"The nuclear industry has changed designs for various plants, but the problems are generic," Oncavage added. "Between Combustion Engineering and Westinghouse, there are twenty-five of these plants in the nation that are suffering from some kind of corrosion, and they haven't been able to solve the problem. Florida Power & Light has added insult to injury by ordering new Westinghouse generators to replace Westinghouse generators." He laughed ironically. "It will get the usual one-year Westinghouse guarantee."*

* Oncavage's figure turned out to be on the conservative side. In March 1982, an N.R.C. staff report stated that steam generator tubes in forty nuclear plants, more than half of those in service, were "virtually impossible" to fix. The report said this defect was responsible for about 23 percent of nuclear plant shutdowns not related to scheduled refueling. (See *The New York Times,* March 31, 1982.)

The licensing board that went to Miami in 1979 decided by a 2-1 vote to give Oncavage a hearing, but this momentary victory, remarkable in itself, didn't change the final decision very much. What Oncavage accomplished, however, was to force FP&L to go back to the drawing board and make some changes. The utility revised its plans for disposing of radioactive material, revised its method of disposing of the coolant, revised its estimate of worker exposure to radiation and changed its method of cutting into the plant to reduce the exposure hazard. These changes persuaded the licensing board to allow FP&L to go ahead.

In its decision, the board cited a staff report that concluded, "A number of changes have been made in the materials, the design, and the operating procedure for replacement of steam generators to assure that the corrosion and denting problems will not recur." The new steam generator design, the report said, "incorporates features that will eliminate the potential for the various forms of tube degradation observed to date." It added, "[It] is *assumed* that the life of the repair is the remainder of the plant life, or about thirty years. *There is no guarantee of this plant life*; however, the Staff safety review found no reason to doubt that steam generators would last the life of the plant." (Italics added.)

One accepts this assumption with some hesitancy since, in the brave new world of nuclear energy, things have rarely gone as planned. Certainly, the Turkey Point repairs have not gone as planned. The licensing board described the power company's proposed course of action.

FPL plans to repair all six generators in Turkey Point Units 3 and 4. The Unit 4 steam generators have the most tubes plugged and, therefore, would be repaired first. The repair of Turkey Point 3 generators is expected to begin about one year later. Since FP&L experiences operating leaks of longer duration in the summer, and the repair is expected to take from six to nine months per unit, the repair should be started in the fall to be completed before the next summer peak demand.

It just didn't work out that way. FP&L, in a footnote to its final brief, explained, "However, because of an unplanned repair outage at Turkey Point Unit No. 3, the repairs were, in fact, commenced on Unit 3 immediately following the authorization of the operating license amendments on June 24, 1981."

Oncavage, who monitored the repair work closely, also contended that Turkey Point No. 4, the newer plant, was the worse plant and should have been dismantled first. But in May a steam generator in No. 3 blew, which meant that the plant would have had to be closed down anyhow. Therefore, FP&L decided to repair that unit right away and to struggle along with No. 4. But then another problem arose. The reactor in No. 4 was suffering from embrittlement. This was in the actual core, and it resulted from the fact that the metal of the core shield had changed its characteristics. The great danger was then posed of a possible thermal shock resulting from a sudden gush of cool water if something should give way. "If this should happen," Oncavage said, "and the core should blow, you can evacuate all of Southern Florida. And the thing is, nobody knows what to do about it."

The embrittlement Oncavage described had taken place in the eight-inch-thick steel shield that surrounds the radio-active core. The N.R.C. has admitted that the constant bombardment of this shield by neutrons from the reactor changes the metal, making it brittle and weak. This has happened not just at Turkey Point but at seven other nuclear plants around the nation. "The pressure vessel of any reactor has to be considered inviolate," said Karl Kniel, a branch chief in N.R.C.'s Division of Safety Technology. "If it were breached, there would be no assurance that we could keep the reactor in a condition that would not lead to a fuel melt-down." And disaster.*

The "thermal shock" Oncavage described actually happened in 1978 at the Rancho Seco nuclear power plant in California. A sudden infusion of cool water caused the temperature inside the pressure vessel to drop 300 degrees Fahrenheit in an hour. Fortunately, this plant had been in operation only three years, and the protective shield had not yet become brittle.

If the entire pressure chamber at Turkey Point No. 4 should have to be replaced, as seems inevitable, it would represent a mammoth task, far beyond the simple dismantling taking place at Turkey Point No. 3. And the costs would be even more astronomical.

Oncavage argued in his brief that energy conservation measures along with solar energy could save more megawatts in sunny Florida than the Turkey Point plants could produce. The hundreds of millions of dollars being spent to recondition the plants, he argued, was "money thrown away." Though the N.R.C. and the licensing board ignored these contentions, they seemed especially relevant after FP&L had to shut down Turkey Point No. 3 during the midsummer peak usage period instead of in the fall and winter as had been planned originally.

---

* Demetrios L. Basdekas, a reactor safety engineer with the N.R.C., wrote on the Op-Ed page of *The New York Times*, March 29, 1982, "There is a high, increasing likelihood that someday soon, during a seeming minor malfunction at any of a dozen or more nuclear plants around the United States, the steel vessel that houses the radioactive core is going to crack like a piece of glass. The result will be a core meltdown, the most serious kind of accident, which will injure many people, destroy the plant, and probably destroy the nuclear industry with it." Basdekas added that this was likely to happen "because the wrong metal was used in the reactor vessels, and with each day of operation, neutron radiation is making the metal core more brittle, and more prone to crack in case of sudden temperature change under pressure."

The additional electricity the utility needed for its 2.2 million customers had to be purchased from plants in southern Georgia at a cost that FP&L estimated at $756,000 a day. Repairs were expected to take 270 days, and when work is started on Turkey Point No. 4, power replacement costs will soar to $809,000 a day. Two-thirds of consumers' bills go to pay for this imported electricity, according to Joel Lumer. As a result, bills of $200 to $300 a month are not uncommon.

The skyrocketing electric bills produced such furious public protest that FP&L went to the extreme of waging a costly television campaign urging more conservation. "Watch Out for Big Bill," the commercials said. There followed the image of a monster labeled "Big Bill" marching across the television screen. Oncavage laughed at the recollection. "There was a backlash," he said. "People didn't have to be told about Big Bill. They already knew about him from the size of their electric bills, and a lot of them got angry."

The legal battle over FP&L's two Turkeys raised other issues. Oncavage argued in his intervention briefs that the availability of other sources of energy and the environmental impact of dismantlement should be considered. The licensing board rejected these arguments, contending that the issues had been settled when the original licenses were granted.

Oncavage demonstrated that the only environmental alternatives originally considered were different sites for the plants and whether their wastes should be discharged into Biscayne Bay or Cord Sound. Conservation measures that might eliminate the need for nuclear power weren't considered, nor was the increased use of solar, a potentially important energy source in Florida, where forty years ago, before electricity became so "cheap," widespread use was made of the sun's rays to heat water.

The narrow interpretation that enabled the licensing

board to dismiss Oncavage's environmental-impact arguments seemed to ignore a vital issue: the problems posed by the disposal of the mass of radioactive wastes to be torn from the two plants. Each of the six steam generators weighs two hundred tons. "They are completely contaminated and radioactive to the core," Oncavage declared.

The power company was going to store them in a building with a dirt floor only 1,300 feet from Biscayne Bay. We fought so hard we won at least one victory. The power company built a hill with the building on top into which the generators were placed on a six- inch concrete floor. [The "hill" is actually a mound seventeen and a half feet high.] Then they added 180 feet of buffer to protect the hill from the waves in Biscayne Bay in case of a storm or hurricane.

So the steam generators were sealed up in this building, but other radioactive wastes were being stored in drums outside the plant. The N.R.C. let me go in and study the situation—the only time, I think, that an outsider has been granted such permission —but then they gagged me. They said I couldn't disclose any of the data I had gathered to anyone.

In one of his last briefs, however, Oncavage entered into the record the fact that 1,312 fifty-five-gallon drums containing wastes had been stored on the site. In addition, he argued, even the N.R.C.'s own studies showed that between 1,100 cubic meters and 2,300 cubic meters of low-level waste, per unit, would be generated by the repairs. This was in addition to the lower assemblies of the steam generators.

When Oncavage carried his fight to the Atomic Safety and Licensing Board in December 1981, he lost his final battle to block the dismantlement of the two Turkeys. However, FP&L countered his waste disposal argument by announcing that it had removed all of the radioactive material Oncavage had found lying around in drums and boxes. Some had been sent to the government's Barnwell, S.C., nuclear-waste disposal facility. Since Barnwell had become so overloaded it had had to cut back on the amount of wastes it would accept, the rest of the radioactive debris had been shipped across the continent to a similar waste-disposal dump in Hanford, Wash.

Such is the story of FP&L's two Turkeys—such the issues raised by a music teacher, horn-blowing David who tackled a utility Goliath. The fact that Oncavage lost his fight isn't as important as the issues he brought to the surface by making it. They have pertinence on a national scale. As Oncavage said in one of his briefs, citing government studies

Fifteen Westinghouse nuclear power units have had adverse experience with their steam generators. . . . Three licensees with Westinghouse units have already filed Steam Generator Repair Reports seeking license amendments to make repairs. . . . There is a good possibility that twelve more applications will be filed. Each of these repair projects will cost hundreds of millions of dollars. In total, the decisions on how to remedy the problem caused by a need for energy and tube degradation in Westinghouse nuclear power units will involve the commitment of billions of dollars.

Those billions will be passed right on to the American consumer. Nuclear power has thus become a multibillion-dollar albatross that utilities can't support, that their stockholders and bondholders can't afford, and that consumers can't bear. Even though complaisant public utility commissions almost invariably impose the costs of big industry's mistakes on the helpless consumer, the problems, have become too immense for the utilities themselves. Hence, the wholesale scrapping of plans to build more nuclear plants and the abandonment of plant after plant on which construction had been started but had been too expensive to finish.

As Mark Oncavage says, "Nuclear power is just not economically feasible." This is the conclusion in other countries besides our own. As the Massachusetts Institute of Technology's *Technology Review* reported, "There is a de facto moratorium on reactor orders in the United States and ten other countries, nuclear energy has been abandoned in at least seven countries, and one country, Sweden, recently voted to phase out nuclear power permanently within twenty-five years."

Whoops!

The Energy Rape

20

Continues

Big Oil is continuing its rape of the American economy. Its actions show that it has every intention of tightening the noose that has almost strangled the nation. Its aim is to maintain high prices—and to drive them ever higher. Its aim is to make the outrageous multibillion-dollar profits of 1979–80 just a springboard for even more astronomical bottom lines. (All that I could learn about the contents of those F.T.C. documents frozen in Representative Dingell's files was that Mobil was projecting an even more glorious and profitable future.) Its aim is to grab for itself ever larger chunks of the American industrial economy.

Anyone who doubts this has only to review Big Oil's consistent record of self-aggrandizement. The phony gasoline "crisis" of 1979 that resulted in the largest profits in Big Oil's history. The almost desperate drive to eliminate natural gas as a competitor and to get its price up to the highest level Big Oil chooses to set. Its piratical raid to force the American public to bear the costs of the Alaska natural gas pipeline to be built for Big Oil's profit. Its grab for another 40 percent of federally owned shale oil lands with no guarantees of development, no guarantees of adequate royalties to the government. Its tapping of the national treasury to help finance its mammoth, multibillion-dollar shale oil and coal gasification projects—white elephants that can be commercially viable only if Big Oil can drive energy prices into the stratosphere—up to levels that seem today, high as prices have been, literally incredible.

In this cause of the ever-higher price, Big Oil has launched its latest ploy—a determined drive supported by some major media like *The New York Times*—to impose a heavy import tax on foreign oil. The proposal is cloaked in the usual propaganda patina of national security and flag-waving, but its real effect would be to put an end to falling prices, reverse the 1982 course and drive prices up again.

The noble endeavor is the result of Big Oil's knee-jerk response to the horrors of 1981–82 that found the world awash in unneeded oil. "The most potent force in the market today," the *Washington Post* noted in mid-March 1982, "is a huge overhang of oil stocks estimated by some experts to be at least 300 million barrels and perhaps as much as 600 million or 700 million worldwide."

It was this "huge overhang" that had brought regular gasoline prices tumbling back to almost the dollar-a-gallon level to which Big Oil had driven them in the 1979 gasoline "crisis." If this situation was allowed to continue, Big Oil would have to forget those monstrous, multibillion-dollar shale oil and coal gasification plants on which it had staked its future and ours.

And so a many-pronged, sophisticated public relations campaign began to creep out of Big Oil executive suites. Look how a a gaggle of synthetic purposes were tied together to impact on the American public. The *Washington Post* noted, "From oilmen suddenly there is talk that perhaps the government should impose an import fee on foreign oil. Advocates say the proceeds from such a fee, which might go as high as $5 or $10 a barrel, could help reduce the large impending federal deficits and encourage continued energy conservation and *synthetic fuels production*." (Italics added.)

How neat. This would solve the budget mess into which Reagan had plunged us with his three-year trillion-dollar defense budget and his annual $100-billion (or more) budget deficits. All this at the same time that he had virtually wiped out our corporate taxes as even David Stockman had confessed in his *Atlantic Monthly* out-of-the-woodshed disclosures. The *Post* noted that the State Department was concerned that "falling oil prices could again put consumption on a rising track," and it quoted a "free market source" within the administration as saying, "The great suspicion here is that the people at State are looking for ways to hold up the price of oil."

The New York Times, in an editorial, "The True Believer, Still," hailed this Big Oil–engineered bit of statesmanship because it would raise revenues through taxes on consumption rather than production(in other words, taxes taken out of the hide of the consuming public rather than from exoribtant Big Oil profits) and because it would encourage beneficial energy savings, making the nation less dependent on foreign oil.

Rep. Silvio O. Conte promptly took the *Times* to task. Even conceding that the oil import fee would raise revenues and that it might encourage further conservation (though he pointed out that conservation had already accomplished almost as much as could be expected of it), he chastised the *Times* for ignoring "the incredible hardship already being caused by petroleum prices."

Conte quoted "one of hundreds of letters" he had received from constituents. It read in part

My take-home pay, after Social Security, state and federal taxes, is $143.95. Our oil bill for the past 13 weeks was $116.25 a week plus utilities, which was $94.26. Gas for our car runs $25 a week. We are not complaining nor are we asking for help.

As Conte pointed out to *The Times*, those numbers just didn't add up. They left nothing for food or clothing. Could a family save enough in summer to compensate for such crushing winter costs? He doubted it. The writer, he said, while not asking for help, had written because he wanted to bring the plight of persons like himself to the attention of public officials. Officials who *just might* do something about it.

Though an import tax would raise revenues, Conte conceded, its impact on the economy would be disastrous. He wrote, "Studies indicate that every dollar of tax per barrel costs consumers $6.8 billion. That is $34 billion of a $5 tax straight out of the hides of consumers."

Stressing that New England is heavily dependent on oil and that its oil costs are the highest in the nation, Conte added, "In 1980, the average homeowner paid $1,300 just in heating costs. A $5-per-barrel import fee would cost the region $2 billion a year. . . . Yes, a fee on imported oil would raise revenues, but it would finish off the already struggling consumer. And yes, it would conserve more energy, while burning out the economy."

It seems strange, if *The New York Times* and the Reagan administration are so concerned about raising revenues, that they don't stop the Big Oil practice of getting its crude supplies from the Middle East and other foreign sources almost for free—at the expense of the government and its other taxpayers.

Twice during the Carter administration, once in 1977 and again in 1979, Representative Rosenthal's Subcommittee on Commerce, Consumer and Monetary Affairs pointed out in communications sent directly to the president in the White House that the nation was losing $1.2 billion a year—and had lost $6 billion between 1974 and 1977—as the result of a tax gimmick Big Oil had negotiated. This was what had happened.

Prior to 1974, major oil companies had been paying Saudi Arabia, for example, a royalty on the crude oil they obtained. The royalty was part of the companies' expense of doing business, but the Saudis thought the royalty was too low. They decided to raise prices drastically on their oil. Big Oil wasn't worried about this. What it said to the Saudis in effect was: "All right, but please call it a tax instead of a royalty." Saudi Arabia didn't have any income tax structure at that time, but it was willing to oblige its Big Oil partners and so it designated the royalty "a tax." Perhaps this doesn't sound like such a big deal to the average American wrestling with a household budget, but a big deal it truly was. By calling its oil purchases "a tax," Big Oil was able to lop off the top of its potential tax bills to Uncle Sam every dollar it had paid to the Saudis and other OPEC nations for their oil. In other words, the government and the American taxpayers by default were buying most of the crude oil supplies for Big Oil's profit.

Even this didn't satisfy Big Oil. There is nothing so beautiful in the world of high finance as making one gimmick give birth to another gimmick. Big Oil did this by establishing refineries in tax-haven countries in the Caribbean. The free, tax-paid oil was sent to these countries which had no tax structure, and the profits from these refineries then flowed into the United States without paying American taxes because those OPEC "taxes" could be applied dollar-for-dollar against any American tax liability.

This complicated double scam was best explained in a hearing before Rosenthal's committee in March 1979. Jack A. Blum, counsel for the Independent Gasoline Marketers Council, was testifying.

Foreign tax credits are generated by the purchase of crude oil from producer countries. Those tax credits are then applied to profits earned in tax haven countries. Normally, if you have profits in a tax haven and attempt to bring them back into the United States, you would be taxed the full U.S. rate. However, when you have surplus tax credits that you have piled up—let's say because you purchased oil in the Middle East—you can apply those tax credits to your profits in the tax haven and bring money back tax free.

Rep. Fernand J. St. Germain wanted to make certain there there was no misunderstanding.

ST. GERMAIN: Let me get this in perspective. A major oil company is purchasing oil in one of the OPEC countries. The price of that oil is considered a tax.

BLUM: That is correct.

ROSENTHAL: The cost of the oil.

ST. GERMAIN: The cost of that oil. It is not taxes paid to the producing country but the cost of the oil itself. . . . They take that oil and go to a tax-haven country and build a refinery where there are no taxes. Is that right?

BLUM: That is correct.

ST. GERMAIN: They then bring the refined product to this country. Now, we are saying we have a shortage of refineries here and we are saying to these same companies, "We are going to give you a tax incentive to build a refinery here to compete with your refinery there."

Blum explained that "willingness to credit goods purchased and treat it as a tax is something that has undermined our energy policy completely. We are in effect saying, 'We will subsidize the high price of OPEC oil.' "

Pennzoil, Exxon, Union Oil and Phillips pulled the same oil-tax deal when they negotiated with the Chinese about developing China's huge oil resources. "Unbelievably enough," Blum testified, "the first thing the companies asked was, 'Will you fellows please pass an income tax so that we can get foreign tax credits on your oil?' "

Blum testified that Carter's D.O.E. (that Big Oil–speckled enclave inside government) had discussed energy policy as if it had no relation to tax policy. "Several times now," he said, "I have brought this matter to the attention of several people in the Department."

I have said, "You must understand what you are doing. You are talking about subsidizing domestic new energy sources. You had better first begin talking about unsubsidizing foreign."

They hired several people who were going to be experts in it. Very quickly, as soon as they started getting out and asking nasty questions, they were transferred to other assignments. They were needed urgently elsewhere.

ROSENTHAL: Why?

BLUM: I have no idea.*

It is not hard to understand how Exxon, for example, chalked up $5.7 *billion* in net income in 1980 and $5.6 *billion* in 1981. That averages out to about $1.4 billion every quarter for the two years—in other words, roughly three times the net income in every quarter that U.S. Steel makes in an entire year. And it was earned in 1981 despite the pressures of the world oil glut, despite falling prices, despite a radical lessening of demand for gasoline (deliveries from primary storage, an important indicator, were down 6.1 percent in the fall from the comparable 1980 period).

Exxon and others among Big Oil kept profits up because they simply did not pass on to consumers the benefits of lower crude oil prices. When OPEC nations met in Geneva, they set a uniform price of $34 a barrel, a move that caused Saudi Arabia to raise prices $2 a barrel while twelve other OPEC nations cut their prices by the same amount. This oversupply and price turmoil had little effect on the American consumer for a long time. The glut caused fluctuating prices in the spot market, and Exxon and some of the other Big Oil companies had their tankers steaming at slow speeds at sea to delay taking deliveries in the hope of striking better bargains. Yet for months, until supplies really surged out of control in early 1982, the oil companies kept gasoline and heating oil prices up.

The Wall Street Journal reported in mid-October 1981, "Major U.S. oil companies have managed to retain most of the benefits from the $2-a-barrel drop in world oil prices in

---

* Rosenthal's committee, after having protested futilely to Carter in 1977, wrote two years later in its follow-up protest: "It is fundamental that IRS will not recognize 'tax avoidance schemes. . . .' Taxpayers with lesser political clout would not be permitted to engage in such a sham arrangement. Nevertheless, we find it incredible that Treasury and IRS have failed to effectively administer Section

recent months, passing on a smaller share of the savings to consumers." The *Journal* estimated that, as a result, third-quarter reports for the major oils were expected to show "earnings equal to or slightly better than the year before, despite a 3 percent to 5 percent drop in consumption."

This prediction was only partially correct. Mobil, Exxon, Texaco and Standard Oil of California reported lower earnings during this supposedly difficult third quarter, but Standard of Indiana's profits rose 36 percent over the booming year of 1980. Standard of Ohio was up 8.4 percent; Gulf, 30 percent; Marathon, 92 percent; Union, 58 percent. Even the companies that reported temporary third quarter setbacks, like Exxon, made up for most of the drop in the fourth quarter and came out of the year rolling in almost as many billions as they had ever known.

This concentration of economic wealth in the hands of a narrowing circle of corporate structures led to an outburst of merger madness—and further concentrations of power. Senator Metzenbaum and Herman Schwartz wrote that, in the first half of 1981, Big Business had spent $35.7 billion in corporate mergers, not far from the $44.3 billion spent in all of 1980. Oil companies, they wrote, "have been particularly voracious. One oil official joked, 'Don't leave anything lying around on the table or we will buy it.' "

It seemed literally true. Mobil offered $5.1 billion for Marathon, and when it was edged out of the bidding by U.S. Steel, a corporation that Marathon preferred, it announced its intention of buying up from 15 to 25 percent of U.S. Steel's stock in a move to gain through leverage in U.S. Steel

---

901 of the Tax Code." This section, as the committee pointed out, would have reduced amounts allowed for foreign tax credits, but it had been ignored in enforcement. And, despite Congress, it has continued to be ignored until recently. Present tax bills, pending in both the Senate and House, would close many such oil company loopholes—and are being vigorously fought by the oil companies.

what it hadn't been able to buy outright in its attempted Marathon takeover.*

Standard Oil of Ohio bought the Kennecott Corporation, the nation's largest copper producer, for $1.77 billion and was ready to spend another $1 billion to modernize Kennecott's facilities. Occidental Petroleum put up $795 million to purchase Iowa Beef Processors because, it said, "food will be in the 1980s what energy has been in the 1970s" and Iowa Beef "is the Cadillac of the industry."

Standard Oil of California, also having some extra billions to scatter around, offered $4 billion for Amax, Incorporated, the largest producer of critically important molybdenum, but was rebuffed by an unsympathetic Amax board. Conoco (before it was taken over by DuPont), Gulf, Occidental, Shell and Standard of Ohio bought up hundreds of thousands of acres containing the largest coal reserves in the nation. Mobil was also in on this action and proclaimed that it was "feeling celebratory, for a change." The reason? Mobil announced that it was developing a "massive" coal mine covering 5,800 acres near Gillette, Wyo., from which it expected to extract 320 million tons of coal a year. It claimed that it also had a method of processing coal-derived methanol into "high-octane gasoline"—a mining, refining and synthesizing process that, it seemed, could be economically feasible only if gasoline prices were kept astronomically high.

Big Oil, as earlier acquisitions had shown, wasn't just interested in acquiring baubles like Montgomery Ward or

* There were many advantages to U.S. Steel in the Marathon takeover. First, it gained access to one of the nation's richest oil reserves, the Yates Field in West Texas. Second, as the results of the favored status given oil companies by the U.S. tax code, it assured itself of another possible bonanza: more than $1 billion in future tax savings. The intricacies of the tax-saving deal remained to be worked out and were subject to interpretation by the Internal Revenue Service. (See *The Wall Street Journal*, March 8, 1982.)

veterinary chemicals or a London newspaper; it was also very big in the highly important agribusiness field. In California's rich Kern County, Shell, Getty, Superior Oil and Tenneco had all become farmers. Tenneco tilled the earth most industriously through its subsidiary, Tenneco West. It had become the largest grower and shipper of table grapes in the nation; it marketed about 25 percent of California almonds; it was the largest shipper of California dates, the third largest producer of raisins; and it processed about 40 percent of the American pistachio harvest.

These were no inconsiderable activities. They seemed to demonstrate that Big Oil had as many billions to spend as any combination of oil-rich shieks. Yet the Reagan administration and the 1981 Congress just couldn't refrain from bestowing upon these poor, suffering oil companies some additional tax relief. Before he left office, Jimmy Carter had persuaded Congress to pass a windfall profits tax to nibble at the excess profits of Big Oil. The tax was expected to yield $227 billion over a ten-year period, but the first Reagan Congress voted to give the oil companies $11.7 billion in "relief" during the first five years. Sen. Edward M. Kennedy, who fought the bill on the Senate floor, contended that it eventually would mean another $33-billion handout for Big Oil.

No wonder the Big Oil companies happened to dominate *Fortune* magazine's list of the top 500 American corporations. In its annual analysis in May 1981, *Fortune* found that thirteen of the top twenty corporations in dollar sales were oil companies. In net income, only International Business Machines (IBM) ranked high, number two behind Exxon. The rest of the top nine were all oil: Mobil, Texaco, Standard of California, Standard of Indiana, Standard of Ohio, Atlantic Richfield and Shell.

TVA Chairman David Freeman had foreseen the time, unless Big Oil were haltered, when a handful of companies would own the nation. It seemed that this time had virtually arrived. It was a development that was being generally ignored in the media, but one which posed critical and fundamental issues for the future of the nation. The Big Oil emphasis on multibillion-dollar shale oil and coal gasification plants, the $40-billion Alaska gas pipeline deal, the drive for natural gas decontrol and an import tax on foreign oil, all could have only one result—driving energy costs into the stratosphere. God help us, let's not find a cheaper source of energy!

But it should be obvious that cheaper energy is precisely what the nation must have if it is ever to break the grip of Big Oil and avoid yet higher inflation and economic chaos. Representative Dingell had noted correctly that Big Oil's exorbitant profits were being piled up at the expense of some nine hundred other American corporations. Aluminum plants, glass factories, textile mills had had to close or cut back operations as rising natural gas prices boosted production costs above a bearable level. Even U.S. Steel and General Motors were feeling the effects of energy rape. Steel was fighting desperately against foreign competition, and the automobile industry, once the bellwether of the national economy, was in a virtual state of collapse. The brutal gasoline prices of 1979–80–81 had driven the American motorist into the arms of Volkswagen, Toyota, Datsun and the rest. Chrysler survived only through the medium of a multibillion-dollar government bailout; Ford had lost $1.54 billion in 1980 and $1.06 billion in 1981. Only General Motors, though badly hurt, was hanging in there.

The airline industry was in shambles. C. Edward Acker, the new chairman of Pan American World Airways, appeared on WNBC-TV's "Live at Five" television show in late March. Acker had taken charge of the huge airline, which had incurred losses ranging into hundreds of millions of dollars. It had had to sell its profitable hotel chain to keep alive, and, even so, even with a $212.6-million gain from the sale, Pan Am had suffered a loss of $18.8 million in 1981. It was now trying to buy South American routes from an airline suffering even worse than itself. Braniff, a once flourishing carrier, was teetering on the edge of bankruptcy and was anxious to obtain federal approval to lease its South American routes to Pan Am for $30 million on a five-year deal. Acker thought the merger of Braniff and Pan Am routes would give his big international carrier the kind of traffic patterns that would result in a profit.*

But why, he was asked, was Pan Am, once one of the largest and most profitable American airlines, in such bad financial shape? He cited instantly the tremendous escalation in the price Pan Am had to pay for kerosene fuel to keep its jets aloft. In 1974, he said, Pan Am was paying 7 cents a gallon; in 1979, 40 cents; in 1981, $1. Fuel costs had become so tremendous they were all but breaking the back of an industry already suffering from the effects of recession.

There is no more excuse for this rape of the airlines through kerosene pricing than there is for the rape of the homeowner through heating oil pricing. The patterns are comparable. As in the case of heating oil, kerosene had always been one of the cheapest products of the refineries (as the 1974 price Acker had quoted showed). Historically, for half a century, it had sold for only 2 or 3 cents a gallon more than No. 2 heating oil—in other words, for less than half the price of refined gasoline at the pump. For a bulk

---

* Pan American failed in its Braniff bid. The Civil Aeronautics Board on April 26, 1982, awarded most of the Braniff routes to Eastern Airlines. But nothing could save Braniff. On May 13, the once profitable carrier filed a petition in bankruptcy. *The Wall Street Journal*'s headline warned: Storm Warnings: The Conditions That Did in Braniff Could Eventually Bring Down Other Airlines as Well. The first quarter 1982 reports gave substance to the warning. Every major American airline was operating in the red. The heaviest losses were posted by United Airlines, $129 million; Pan American, $127 million; and Trans World, $102 million.

purchaser like Pan Am to be forced to pay a dollar a gallon, a price comparable to what the individual driver was paying for gasoline, is just another example of the colossal rip-off that is ruining the American economy.

Powerful American industries with their financial resources should have been able to muster the kind of clout needed to combat Big Oil. Instead, they have acquiesced in the banditry. The Bank of America was the only large institution, as far as I could discover, that had rebelled against the 1979 holdup and had tried to do something about it by using methanol. The rest of American big business was like a helpless chicken with its neck stretched out on the chopping block waiting for the ax to fall.

It is understandable only in terms of ideology. Big business has indoctrinated itself with the myth of the wonders of a functioning free-enterprise system. It shudders at the idea of government regulation and control. Even when the gun is being poked into its own ribs, it still has horrors at the idea that if government interferes with Big Oil it may interfere with others tomorrow. And so it rolls over and plays dead. It is its own victim.

This will not change unless business and America as a whole get away from the idea that bigness is better. It just isn't. Senator Kennedy pointed this out in a letter to *The New York Times*.

Numerous independent analyses, including studies by the National Science Foundation [N.S.F.], the Massachusetts Institute of Technology, the Office of Management and the Budget and the Department of Commerce, have concluded that smaller firms, with their propensity for risk-taking and fresh ideas, are the leading sources of innovation. The N.S.F. study found that they have developed nearly half of all major innovations since World War II.

Kennedy was defending a measure he had introduced to allocate a tiny portion of the research budget—only some

$50 million a year—to smaller firms to promote basic research. He was supported in the House by Rep. John J. LaFalce, head of the Business Committee's subcommittee on oversight. LaFalce emphasized that the National Science Foundation study showed that "small firms produce twenty-four times as many major innovations per R & D dollar as large firms, and four times as many as medium-sized firms."

Kennedy and LaFalce were speaking of the whole range of research and development, not specifically addressing the energy issue, but what they had to say applies. To break the Big Oil monopoly on energy, we need the kind of innovation that brought the world Edison's electric light bulb, the Wright brothers' airplane, Ford's Model T.

Much more can be done in many energy areas to give the nation some viable alternatives and, perhaps, what is needed most—cheaper, domestic sources of power that do not depend on the price manipulations of OPEC and Big Oil. Except for the farmers who see the possibility of energy independence in ethanol, there is little determined effort along these lines. Solar and the photovoltaic cell offer a wide range of possibilities: heating, electrical generation, even the splitting of the hydrogen atom from water to make a cheap, safe hydrogen- hydride motor fuel.

As far back as August 1977, Vincent Esposito, then director of the Division of Transportation and Energy Conservation in the Energy Research and Development Administration (the scientific arm of D.O.E.) wrote to a correspondent of mine.

Hydrogen is an excellent motor fuel. It has been identified as a long-term candidate alternative fuel because its use results in very low emissions. However, it is not likely to be available in quantity at an attractive price until we have sufficient energy (e.g., nuclear or solar) to make hydrogen from water.

Esposito added that hydrogen fuel could not become practical until an entirely new engine had been designed to accommodate it—and that, he estimated, would postpone its use to the year 2000. In other words, at least two more decades of grace for Big Oil and its enormous profits.

Big Oil can be expected to do its utmost to kill off such a dangerous competitor at birth. The signs, indeed, are clear. Among the first projects eliminated by the U.S. Synthetic Fuels Corporation were those involving solar and hydrogen. The corporation was following the Big Oil script that calls for shale oil and coal gasification—the ruinous roads to a higher-level price plateau.

In winding up this account of my three-year investigation of energy problems, I find that I am left with one dominant conclusion: truth is simply not to be found in the higher echelons of American government and big business. Spokesmen for the oil industry lied on virtually every major issue. They lied when they insisted their prices were purely cost-related—only to have to admit to congressional investigators that their only principle was to charge whatever the traffic would bear; they lied in insisting that the "Iranian shortfall" had caused the 1979 gasoline "crisis." Government under Jimmy Carter was no better. D.O.E. lied repeatedly in asserting it was monitoring heating oil prices when, as it later had to admit to congressional staffs, it was not monitoring them at all; D.O.E. and the Justice Department deceived the public in their transparently phony whitewashes of Big Oil's role in the 1979 "crisis." And what about Jimmy Carter himself? He occasionally grabbed headlines by jawboning about the iniquities of Big Oil, but the record shows that anyone in his administration who opposed Big Oil interests was either discredited or ousted. Despite all the evidence to the contrary—some of it delivered to him by his closest aides—he kept insisting that the 1979 "crisis" was real, and he put out under the imprimatur of the White House the sleazy rationalizations of D.O.E. and the Justice Department.

The victims were the American people—and, though it is not generally recognized, the vaunted American free-enterprise system. Ideologically, this system may indeed be a great one. If it exists. The fact is that it does not. I have shown here, I believe, ample evidence of Big Oil's control of the machinery of government and the American economy. It is the kind of control that could not be exerted in a truly competitive, free-enterprise system—only in an ogopolistic, monopolistic order where a relative handful of super-powerful men determine the fate of the nation.

This is not just my own conclusion. Four-star Admiral Hyman G. Rickover, the father of the nuclear navy, addressed the Joint Economic Committee of Congress after retiring on January 31, 1982, after Reagan became president. Admiral Rickover had seen in naval procurement the same operation of powerful forces, the same immunity to prosecution, that I had encountered in my energy investigation. In his blunt and forceful way, he demolished the myth that we have a competitive free-enterprise society. He made it clear that he thought our free-enterprise system the greatest that had ever been devised if it was allowed to work. The trouble was that it had been corrupted and distorted beyond recognition.

"A preoccupation with the so-called bottom line of profit-and-loss statements, coupled with a lust for expansion, is creating an environment in which fewer businesspeople honor traditional values . . . ," he said.

"Political and economic power is increasingly being concentrated among a few large corporations and their officers —power they can apply against society, government and individuals. Through their control of vast resources, these large corporations have become, in effect, another branch of government. They often exercise the power of government, but without the checks and balances inherent in our democratic system. . . .

"The notion that we are a self-regulating, free-market economy that prompts a high standard of ethical business conduct is not realistic in today's complex society. . . ."

This preoccupation with the "bottom line" without regard for ethical principles, without any consideration for the damage being done to the nation, has produced a system whose only god is power and the wealth that goes with it. Heating oil prices soared so high that hypothermia put thousands of the elderly at risk and caused an uncounted number of deaths, yet the circumstance, if it was noted at all, rated only a casual line or two in the press. So much for human life. We have created a political and economic system that is the antithesis of that envisioned by our Revolutionary forebears—a society wedded to big money, to big bucks solutions that don't always solve (as in nuclear power and shale oil); a society that reduces the average American to a nonentity—unless the time should come, as it may, when he will be needed to go to the Persian Gulf and fight for Exxon and the rest of the Seven Sisters.

Index